Deep Learning with fastai Cookbook

Leverage the easy-to-use fastai framework
to unlock the power of deep learning

Mark Ryan

BIRMINGHAM—MUMBAI

Deep Learning with fastai Cookbook

Group Product Manager: Kunal Parikh

Publishing Product Manager: Ali Abidi

Senior Editor: Roshan Kumar

Content Development Editor: Tazeen Shaikh

Technical Editor: Sonam Pandey

Copy Editor: Safis Editing

Project Coordinator: Aparna Ravikumar Nair

Proofreader: Safis Editing

Indexer: Manju Arasan

Production Designer: Vijay Kamble

First published: September 2021

Production reference: 2260821

Published by Packt Publishing Ltd.
Livery Place
35 Livery Street
Birmingham
B3 2PB, UK.

ISBN 978-1-80020-810-0

www.packt.com

To the memory of my father, Ben. He would have loved fastai.

– Mark Ryan

Contributors

About the author

Mark Ryan is a machine learning practitioner and technology manager who is passionate about delivering end-to-end deep learning applications that solve real-world problems. Mark has worked on deep learning projects that incorporate a variety of related technologies, including Rasa chatbots, web applications, and messenger platforms. As a strong believer in democratizing technology, Mark advocates for Keras and fastai as accessible frameworks that open up deep learning to non-specialists. Mark has a degree in computer science from the University of Waterloo and a Master of Science degree in computer science from the University of Toronto.

I want to thank my family for their support during the development of this book. In particular, I would like to acknowledge my nephew, Rowan Hansen, for his advice on web development. My friends, in particular Dr. Laurence Mussio, Peter Moroney, Luc Chamberland, and Alan Hall, provided much-appreciated support to me while I wrote this book. Finally, I would like to thank the team at Packt for guiding me through the process of completing this book.

About the reviewer

Rupsi Kaushik is a backend engineer at the Paris-based AI start-up reciTAL.
She graduated from the University of Ottawa with a B.Sc. in computer science with an
option in entrepreneurship and management. Her interest in machine learning was first
sparked when she created her first search engine from scratch, and it has grown ever since.
She hopes to learn more about AI for social good and make a meaningful impact on
the world.

Table of Contents

2

Exploring and Cleaning Up Data with fastai

3

Training Models with Tabular Data

Preface

fastai is an easy-to-use deep learning framework built on top of PyTorch that lets you rapidly create complete deep learning solutions with as few as 10 lines of code. Both of the predominant low-level deep learning frameworks today (TensorFlow and PyTorch) require a lot of code, even for straightforward applications. In contrast, fastai handles the messy details for you and lets you focus on applying deep learning to actually solve problems.

We will start by summarizing the value of fastai and showing a simple "hello world" deep learning application with fastai. Then, we will describe how to use fastai for each of the four application areas that the framework explicitly supports: tabular data, text data (NLP), recommender systems, and vision data. You will work through a series of practical examples that illustrate how to create real-world applications of each type. After that, you will learn how to deploy fastai models. For example, you will learn how to create a simple web application that predicts what object is depicted in an image. Finally, we will wrap up with an overview of the advanced features of fastai.

By the end of this book, you will be able to create your own deep learning applications using fastai. You'll know how to use fastai to prepare raw datasets, explore datasets, train deep learning models, and deploy trained models.

Who this book is for

This book is for data scientists, machine learning developers, and deep learning enthusiasts who are looking to learn and explore the fastai framework using a recipe-based approach. Working knowledge of the Python programming language and machine learning basics are strongly recommended to get the most out of this book.

This book provides practical examples of how to use fastai to tackle a variety of deep learning application areas, but it is not an exhaustive reference for the platform. To get comprehensive details on fastai, please see the *Conclusion and additional resources on fastai* section in *Chapter 8, Extended fastai and Deployment Features*. This section points to additional fastai content, including the excellent deep learning courses built around fastai created by Jeremy Howard and his team.

What this book covers

Chapter 1, Getting Started with fastai, shows you how to set up an environment for fastai, takes you through training an initial *hello world* fastai model, explains the four key application areas of fastai (tabular data, text data, recommender systems, and image data), and compares fastai with the other important high-level deep learning framework, Keras.

Chapter 2, Exploring and Cleaning Up Data with fastai, describes the set of datasets that fastai makes available out of the box (the **curated** datasets); describes how to examine tabular, text, and image datasets; and shows how to use the facilities of fastai to clean up a dataset, for example, by dealing with missing values.

Chapter 3, Training Models with Tabular Data, explains how to create fastai deep learning models trained on tabular datasets, that is, datasets that are arranged in rows and columns. Examples in this chapter show you how to train fastai models on both curated and standalone datasets.

Chapter 4, Training Models with Text Data, explains how to create fastai deep learning models trained on text datasets. Examples in this chapter show you how to train language models (that is, models that predict the next word given a series of words), as well as how to train text classification models (that is, models that predict, for example, whether a given review is negative or positive). This chapter covers models trained with both curated and standalone datasets.

Chapter 5, Training Recommender Systems, explains how to use fastai to create recommender systems, that is models that predict, for example, whether a particular reader will like a particular book given a set of ratings that other readers have provided for this book. This chapter covers recommender systems trained with both curated and standalone datasets.

Chapter 6, Training Models with Visual Data, explains how to use fastai to create deep learning models trained on image datasets. Examples in this chapter show you how to create image classification systems for images that depict one or more objects, trained on both curated and standalone datasets.

Chapter 7, Deployment and Model Maintenance, explains how you can take a trained fastai model and deploy it in a simple web application. The examples in this chapter show you how to deploy fastai deep learning models trained on tabular and image datasets. This chapter also tells you how to maintain models once they have been deployed.

Chapter 8, Extended fastai and Deployment Features, explains additional aspects of fastai, including enhancements to the models introduced from *Chapter 3, Training Models with Tabular Data,* to *Chapter 6, Training Models with Visual Data,* as well as variations on the deployment techniques introduced in *Chapter 7, Deployment and Model Maintenance.*

To get the most out of this book

To get the most out of this book, you should be comfortable with coding in Python (in Jupyter notebooks and in standalone Python modules) and with the core concepts of machine learning. This book explains a broad variety of deep learning applications but doesn't go into the internals of deep learning itself. If you have a basic grasp of how deep learning works, you will find the more advanced examples in the book easier to follow.

Software/hardware covered in the book	OS requirements
Python 3.7	Windows or Linux
Python libraries: pandas, Folium	Windows or Linux
Jupyter notebook	Windows or Linux
Cloud deep learning environment: Paperspace Gradient, Google Colaboratory	Windows or Linux
Deep learning frameworks: fastai, PyTorch, Keras	Windows or Linux

Most of the code examples in this book are designed to be run in GPU-enabled cloud deep learning Jupyter notebook environments. You have the choice of using either **Paperspace Gradient** or **Google Colab** for these examples, with Gradient being the recommended environment. The model deployment examples in *Chapter 7, Deployment and Model Maintenance*, and *Chapter 8, Extended fastai and Deployment Features*, are designed to be run on your local system and require fastai and PyTorch to be installed on your local system.

If you are using the digital version of this book, we advise you to type the code yourself or access the code via the GitHub repository (link available in the next section). Doing so will help you avoid any potential errors related to the copying and pasting of code.

Download the example code files

You can download the example code files for this book from GitHub at `https://github.com/PacktPublishing/Deep-Learning-with-fastai-Cookbook` In case there's an update to the code, it will be updated on the existing GitHub repository.

We also have other code bundles from our rich catalog of books and videos available at `https://github.com/PacktPublishing/`. Check them out!

Download the color images

We also provide a PDF file that has color images of the screenshots/diagrams used in this book. You can download it here: `https://static.packt-cdn.com/downloads/9781800208100_ColorImages.pdf`.

Conventions used

There are a number of text conventions used throughout this book.

`Code in text`: Indicates code words in text, database table names, folder names, filenames, file extensions, pathnames, dummy URLs, user input, and Twitter handles. Here is an example: "Go to `localhost:5000` in your browser to display `home.html`."

A block of code is set as follows:

```
for(var i = 0; i < relationship_list.length; i++) {
            var opt = relationship_list[i];
            select_relationship.innerHTML += "<option value=\""
+ opt + "\">" + opt + "</option>";
```

Any command-line input or output is written as follows:

```
cp -r deploy_image deploy_image_test
```

Bold: Indicates a new term, an important word, or words that you see onscreen. For example, words in menus or dialog boxes appear in the text like this. Here is an example: "Select the **Choose Files** button to bring up the file selection dialog."

> **Tips or important notes**
> Appear like this.

Sections

In this book, you will find several headings that appear frequently (*Getting ready, How to do it..., How it works..., There's more...,* and See also).

To give clear instructions on how to complete a recipe, use these sections as follows:

Getting ready

This section tells you what to expect in the recipe and describes how to set up any software or any preliminary settings required for the recipe.

How to do it...

This section contains the steps required to follow the recipe.

How it works...

This section usually consists of a detailed explanation of what happened in the previous section.

There's more...

This section consists of additional information about the recipe in order to make you more knowledgeable about the recipe.

See also

This section provides helpful links to other useful information for the recipe.

Get in touch

Feedback from our readers is always welcome.

General feedback: If you have questions about any aspect of this book, mention the book title in the subject of your message and email us at `customercare@packtpub.com`.

Errata: Although we have taken every care to ensure the accuracy of our content, mistakes do happen. If you have found a mistake in this book, we would be grateful if you would report this to us. Please visit `www.packtpub.com/support/errata`, selecting your book, clicking on the Errata Submission Form link, and entering the details.

Piracy: If you come across any illegal copies of our works in any form on the Internet, we would be grateful if you would provide us with the location address or website name. Please contact us at `copyright@packt.com` with a link to the material.

If you are interested in becoming an author: If there is a topic that you have expertise in and you are interested in either writing or contributing to a book, please visit `authors.packtpub.com`.

Share Your Thoughts

Once you've read *Deep Learning with fastai Cookbook*, we'd love to hear your thoughts! Scan the QR code below to go straight to the Amazon review page for this book and share your feedback.

`https://packt.link/r/1-800-20810-3`

Your review is important to us and the tech community and will help us make sure we're delivering excellent quality content.

1
Getting Started with fastai

Over the last decade, deep learning has revolutionized swathes of technology, from image recognition to machine translation. Until recently, only those with extensive training and access to specialized hardware have been able to unlock the benefits of deep learning. The fastai framework is an effort to democratize deep learning by making it accessible to non-specialists. One of the key ways that fastai opens up deep learning to the masses is by making it easy to get started.

In this chapter, we will show you what you need to get started with fastai, starting with how to set up an environment for fastai. By the end of this chapter, you will be able to do the following: set up a cloud environment in which to run `fastai` examples; exercise a basic fastai example; explain the relationship between fastai and PyTorch (the underlying deep learning library for fastai); and contrast fastai with Keras, the other high-level library for deep learning.

Here are the recipes that will be covered in this chapter:

- Setting up a fastai environment in Paperspace Gradient
- Setting up a fastai environment in Google Colaboratory (Google Colab)
- Setting up JupyterLab environment in Paperspace Gradient

- "Hello world" for fastai—creating a model for the **Modified National Institute of Science and Technology (MNIST) dataset**
- Understanding the world in four applications: tables, text, recommender systems, and images
- Working with PyTorch tensors
- Contrasting fastai with Keras
- Test your knowledge

Technical requirements

For this chapter, you will be using the following technologies:

- Paperspace Gradient: `https://gradient.paperspace.com/`
- Google Colab: `https://colab.research.google.com/notebooks/intro.ipynb`
- Google Drive: `https://drive.google.com`
- Keras: `https://keras.io/`

You can find the code referred to in this chapter at the following link:

`https://github.com/PacktPublishing/Deep-Learning-with-fastai-Cookbook/tree/main/ch1`

Setting up a fastai environment in Paperspace Gradient

There are two free cloud environments that you can use to explore fastai: **Paperspace Gradient** and **Google Colab**. In this section, we'll go through the steps to set up Paperspace Gradient with a fastai notebook environment, and in the next section, we'll go through the setup steps for Colab. It's your choice, so pick the environment that works best for you.

Gradient is simpler to use because you have access to a standard filesystem for storage. With Colab, you need to use Google Drive for storage and, unlike Gradient, you don't have convenient access to the terminal for command-line interactions.

On the other hand, Colab gives you direct access to a wider set of libraries beyond those needed for fastai—for example, you can run the Keras MNIST example in Colab but it won't work off the shelf in a Gradient fastai instance. To get the most out of the examples in the book, it's best to set up both environments so that you can choose which one works best for you as you go along. We'll start with Gradient, since it is the simplest to get started with.

Getting ready

Prior to setting up Gradient for fastai, you need to create a Paperspace account. You can do this by going to `https://console.paperspace.com/signup?gradient=true`.

How to do it...

Once you have a Paperspace account, you can create a free fastai notebook in Gradient by following these steps to create a fastai notebook instance in Gradient. Once created, this will be a complete Jupyter Notebook environment with all the libraries that you need (including fastai, PyTorch, and related libraries).

1. Go to the Paperspace site and sign in using the account you created in the *Getting ready* section.

2. From the pulldown at the top of the page, select **Gradient**:

Figure 1.1. – Selecting gradient from the pulldown

3. Select the **Notebooks** tab:

Figure 1.2 – Select the Notebooks tab

4. Select the **CREATE** button.

Figure 1.3 – CREATE button

5. Enter a name for your notebook in the **Name** field.

6. In the **Select a runtime** section, select **fastai**.

7. In the **Select a machine** section, select **Free-GPU** or **Free-P5000**. Note that you may receive a message indicating out of capacity for the machine type you selected. If this happens, you can either choose another GPU-enabled machine type or wait a few minutes and try again with your original machine type. Also note that after your notebook is created, you can change the machine type—for example, if you find that the free instance is not meeting your needs, you can switch your notebook to a paid machine. You can also define multiple notebooks for different applications and configure auto-shutdown (how many hours your instance will run before shutting itself down) if you opt for a paid subscription. For details, see `https://console.paperspace.com/teim6pi2i/upgrade`.

8. Select the **START NOTEBOOK** button to launch the process of creating a new fastai instance for you in Gradient.

Figure 1.4 – START NOTEBOOK button

9. Your notebook will take a minute or so to be created. When it is ready, you will see a **Running** message at the bottom of the screen:

Figure 1.5 – Running message

10. Next, you should see a Jupyter button appear in the navigation panel on the left, as highlighted in *Figure 1.6*:

Figure 1.6 – Jupyter icon in the navigation panel

11. Select the Jupyter button to start your new notebook environment. You should now see a Jupyter files view, as shown in *Figure 1.7*:

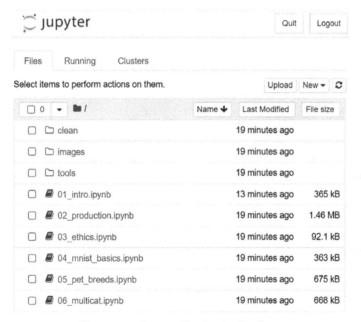

Figure 1.7 – Jupyter file view in Gradient

Now that your notebook has started, you need to validate that it was set up correctly by running a short notebook that checks the fastai version available to your notebook and confirms that your notebook has access to **graphics processing units** (**GPUs**), the specialized hardware required to efficiently run subsequent examples in this book:

1. Open a terminal in the root directory of your Gradient notebook environment:

Figure 1.8– Pulldown to open a terminal in Jupyter notebook

2. In the terminal, create a new directory, `fastai_cookbook`, in the root level of your notebook:

    ```
    mkdir fastai_cookbook
    ```

3. In the terminal, make this new directory your current directory:

    ```
    cd fastai_cookbook
    ```

4. Initialize `git` in this new directory:

    ```
    git init
    ```

5. Clone the repository for the book:

    ```
    git clone https://github.com/PacktPublishing/Deep-
    Learning-with-fastai-Cookbook.git
    ```

6. Once the repository has been cloned, go to the ch1 directory and open the
    validate_gradient_setup.ipynb notebook:

Figure 1.9 – validate_gradient_setup.ipynb notebook in the Files view

7. Run the entire notebook (**Cell -> Run all**) and check the output.

8. For the first code cell, you should see something like the following if your notebook
    has access to the fastai library. Don't worry about the exact level of fastai—the
    key point is that you are able to import the library and get back a valid version
    without errors:

```
# validate the version of fast.ai
import fastai
fastai.__version__

'2.1.5'
```

Figure 1.10 – Getting the fastai version

9. For the second code cell, you should see something like the table shown next if your
    notebook has access to a GPU. A GPU is specialized hardware for deep learning that
    you will need in order to efficiently run subsequent examples. Don't worry about
    the specific type of GPU listed; just confirm that you get a table like this as output
    of this cell:

```
# validate access to GPUs
!nvidia-smi

Sun Dec 27 18:12:30 2020
+-----------------------------------------------------------------------------+
| NVIDIA-SMI 450.36.06    Driver Version: 450.36.06    CUDA Version: 11.0      |
|-------------------------------+----------------------+----------------------+
| GPU  Name        Persistence-M| Bus-Id        Disp.A | Volatile Uncorr. ECC |
| Fan  Temp  Perf  Pwr:Usage/Cap|         Memory-Usage | GPU-Util  Compute M. |
|                               |                      |               MIG M. |
|===============================+======================+======================|
|   0  Quadro P5000        On   | 00000000:00:05.0 Off |                  Off |
| 26%   20C    P8     6W / 180W |    1MiB / 16278MiB   |      0%      Default |
|                               |                      |                  N/A |
+-------------------------------+----------------------+----------------------+
```

Figure 1.11 – Output of the nvidia-smi command

If you get the following kind of output from this cell, then your notebook was not set up correctly with access to a GPU:

```
!nvidia-smi

NVIDIA-SMI has failed because it couldn't communicate with the NVIDIA driver.
```

Figure 1.12 – Error from the nvidia-smi command

Congratulations! You have set up a Gradient environment that is ready to explore fastai.

How it works...

Now that you have a working Gradient instance, you will be able to run fastai examples. Gradient includes PyTorch, fastai, and other libraries that you need to run the examples in this book, along with access to the GPU hardware that you need to run these examples efficiently.

Some of the aspects of Gradient notebooks that you need to be aware of are listed here:

- By default, your free instance will run for 6 hours and then shut itself down. If you want to have longer, uninterrupted sessions, you will need to change to a paid subscription.

- Generally speaking, restarting a Gradient instance takes between 3 and 10 minutes, so it's a good idea to go to the **Notebook** section of the Paperspace console and click on **START** for your notebook a few minutes before you're ready to actually get working. I am in the habit of starting my notebook and then completing some other task (such as sending an email or making a cup of tea) so that I'm not waiting too long for the notebook to start.

- If you are a bit rusty about how to use Jupyter notebooks, the tutorial available at https://www.dataquest.io/blog/jupyter-notebook-tutorial/ gives a good review of the key points.

There's more...

If you have completed all the steps in this section and have a working Gradient environment, the next section is not strictly required. I recommend that you set up both Gradient and Colab, but it's not mandatory to have both environments in order to complete most of the examples in this book. However, if you want the best of both worlds, you can also set up Colab for fastai—it's also free, and it offers some advantages over Gradient, such as supporting Keras applications.

Setting up a fastai environment in Google Colab

If you are already familiar with the **Google Colab** environment or want to take advantage of Google's overall machine learning ecosystem, Colab may be the right environment for you to use to explore fastai. In this section, we'll go through the steps to get set up with Colab and validate that it's ready for you to use with fastai.

Getting ready

To use Colab, you will need a Google ID and access to Google Drive. If you don't already have a Google ID, follow the instructions here to create one: `https://support.google.com/accounts/answer/27441?hl=en`.

Once you have a Google ID, you need to confirm that you have access to Google Drive. You need access to Drive because it acts as the storage system for Colab. You save your notebooks and data in Drive when you are working in Colab. Follow the instructions here to get access to Drive: `https://support.google.com/drive/answer/2424384?co=GENIE.Platform%3DDesktop&hl=en`.

How to do it...

Once you have a Google ID with access to Drive, you can set up Colab to work with fastai by completing the following steps. First, we'll get access to Drive from within a Colab notebook, then clone the repository for this book, and finally run the validation notebook to confirm the setup worked.

1. Open Colab (`https://colab.research.google.com/`).
2. Open a new, blank notebook by selecting **File -> New notebook**.
3. In the new notebook, paste the following statement into an empty cell:

    ```
    print("hello world")
    ```

    Then, select the **Run** button:

Figure 1.13 – Colab run button

4. Confirm that you get the expected output:

Figure 1.14 – Expected output of "hello world" in Colab

5. Go to Drive and create a new folder called `fastai_cookbook` in your root folder in Drive.

6. Go into this new folder and right-click, and select **Google Colaboratory**:

Figure 1.15 – Selecting Google Colaboratory in your new directory in Drive

7. Colab will open a new notebook. In this notebook, select **Connect** -> **Connect to hosted runtime**:

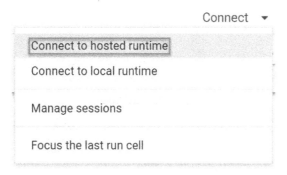

Figure 1.16 – Selecting Connect to hosted runtime

8. In a new cell in this notebook, paste the following code and run the cell (for example, by clicking the arrow):

```
from google.colab import drive
drive.mount('/content/drive')
```

9. In the response that comes back, click on the link that is provided:

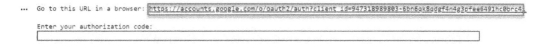

Figure 1.17 – Prompt to mount Google Drive in your notebook

10. Select an account:

Sign in with Google

Choose an account

to continue to Google Drive File Stream

Mark Ryan
███████████@gmail.com

Use another account

To continue, Google will share your name, email address,
language preference, and profile picture with Google Drive
File Stream. Before using this app, you can review Google
Drive File Stream's **privacy policy** and **terms of service**.

Figure 1.18 – Dialog to select your Google account

11. On the screen for **Google Drive File Stream** access, select on **Allow**:

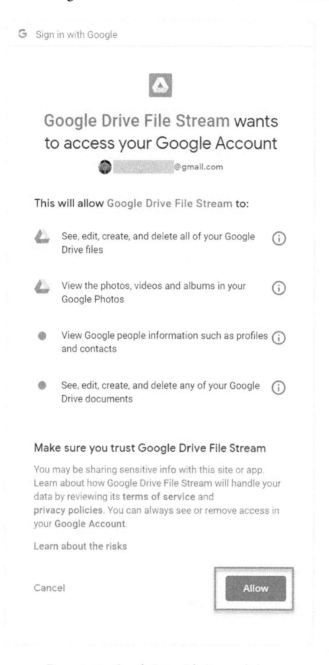

Figure 1.19 – Google Drive File Stream dialog

12. On the **Sign in** screen, select the **copy** icon to copy your access code:

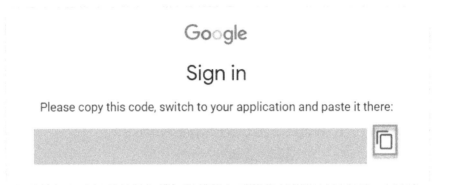

Figure 1.20 – Dialog to get access code

13. Now, return to the notebook in Colab and paste the access code in the authorization code field, and then press *Enter*:

```
from google.colab import drive
drive.mount('/content/drive')
```

```
Go to this URL in a browser: https://accounts.google.com/o/oauth2/auth?client_id=947318989803-6bn6qk8qdgf4n4g3pfee6491hc0brc4i
```

```
Enter your authorization code:
•••••••••••••••••••••••••••••••••••••••••••••••••••••
```

Figure 1.21 – Access code entered to mount Google Drive

14. The cell will run and produce the following mounted message to confirm that your Google Drive has been mounted and is available for your Colab notebook:

```
from google.colab import drive
drive.mount('/content/drive')
```

```
Go to this URL in a browser: https://accounts.google.com/o/oauth2/auth?client_id=947318989803-6bn6qk8qdgf4n4g3pfee6491hc0brc4i
```

```
Enter your authorization code:
..........
Mounted at /content/drive
```

Figure 1.22 – Message confirming that Google Drive has been mounted

Now that Drive is mounted in Colab, the next step is to clone the book's repository:

1. Make the `fastai_cookbook` new directory folder in Drive your current directory by running a cell in the notebook with the following command:

```
%cd /content/drive/MyDrive/fastai_cookbook
```

2. Run the following command in a new cell to list the contents of this directory:

```
%ls
```

3. Run the following code in a new cell in your notebook to clone the book's repository:

```
!git clone https://github.com/PacktPublishing/Deep-
Learning-with-fastai-Cookbook.git
```

4. Run the cell to list the directory contents again, and you should see the directory created now for the repository. You can confirm in Drive that the repository has been cloned.

Now that you have cloned the repository, you can run the validation notebook to confirm that you have access to the `fastai` library and GPUs:

1. In Drive, navigate to the `fastai_cookbook/Deep-Learning-with-fastai-Cookbook/ch1` folder, right-click on the `validate_gradient_setup.ipynb` notebook, and select **Open With** | **Google Colaboratory**.

2. The notebook opens up in Colab. Select **Runtime** | **Change Runtime Type**. In the **Notebook settings** dialog that comes up, select **GPU** in the **Hardware accelerator** field, and select **SAVE**:

Figure 1.23 – Selecting GPU as the hardware accelerator in the Notebook settings dialog

3. Run the notebook by selecting **Runtime** | **Run all**.

4. Confirm that you get output like the following, with no errors, for the first code cell in the notebook:

```
# validate the version of fast.ai
import fastai
fastai.__version__

'1.0.61'
```

Figure 1.24 – Confirmation of the fastai version

5. Confirm that you get output like the following for the second code cell in the notebook. Don't worry about the specific GPU type listed—this will vary depending on what's available. If you did not specify **GPU** as the hardware accelerator in *Step 2*, then you won't get this output:

```
# validate access to GPUs
!nvidia-smi

Mon Dec 28 03:36:32 2020
+-----------------------------------------------------------------------------+
| NVIDIA-SMI 460.27.04    Driver Version: 418.67       CUDA Version: 10.1     |
|-------------------------------+----------------------+----------------------+
| GPU  Name        Persistence-M| Bus-Id        Disp.A | Volatile Uncorr. ECC |
| Fan  Temp  Perf  Pwr:Usage/Cap|         Memory-Usage | GPU-Util  Compute M. |
|                               |                      |               MIG M. |
|===============================+======================+======================|
|   0  Tesla T4            Off  | 00000000:00:04.0 Off |                    0 |
| N/A   39C    P8    10W /  70W |      0MiB / 15079MiB |      0%      Default |
|                               |                      |                 ERR! |
+-------------------------------+----------------------+----------------------+
```

Figure 1.25 – Output of nvidia-smi confirming access to a GPU

Congratulations! You have set up a Colab environment that is ready to explore fastai.

How it works...

Now that you have a working Colab environment, you will be able to run fastai examples in it. Colab incorporates PyTorch, fastai, and other libraries that you need to run the examples in this book. Note that, unlike Gradient, every time you start up a new Colab session, you will need to follow the steps to mount Drive and will also need to specify that you want a GPU. By default, Drive is not mounted and your Colab notebooks don't have access to GPUs until you explicitly change the hardware accelerator type.

There's more...

If you have set up both Gradient and Colab environments, I recommend that you use Gradient to exercise the examples in this book by default. Gradient gives you direct access to a terminal, which is handy for entering command-line commands, and does not require you to mount a filesystem or request a GPU every time you start a new session. Colab does have some advantages, including not shutting down after 6 hours, but overall you will have a smoother experience with Gradient.

Setting up JupyterLab environment in Gradient

Earlier in this chapter, we went through the steps to set up Gradient as an environment to explore fastai. With this set up, you get the standard Jupyter notebook environment that features a filesystem view and the ability to update notebooks, launch terminal windows, and perform basic operations such as uploading and downloading files from your local system. If you want a richer development environment, you can set up Gradient to use JupyterLab.

In addition to allowing you to maintain multiple views (for example, a terminal view along with several notebooks) within the same browser tab, JupyterLab also lets you take advantage of visual debuggers in the context of a notebook. In this section, we will go through the steps to set up Gradient so that you can use JupyterLab. Note that this recipe is optional—any example in this book that you can run in Gradient with JupyterLab will also work in vanilla Jupyter.

Getting ready

Before you attempt to set up Gradient with JupyterLab, ensure that you have successfully completed the steps in the *Setting up a fastai environment in Paperspace Gradient* section. Once you have set up JupyterLab Gradient, you will be able to switch back and forth between the vanilla Jupyter view and JupyterLab at any time.

How to do it...

To get JupyterLab set up, you begin by starting up your Gradient instance, run a command to install JupyterLab, and then restart the instance to see the result. Here are the steps to do this:

1. Start your Gradient fastai instance to bring up the vanilla Jupyter **Files** view:

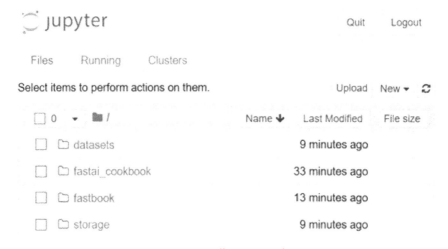

Figure 1.26 – Vanilla Jupyter Files view

2. Once you are in the filesystem view for your instance, select **New | Terminal**:

Figure 1.27 – Pulldown to open up a terminal in Jupyter

3. This will open a terminal window:

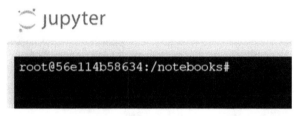

Figure 1.28 – Jupyter terminal window

4. In the terminal window, enter this command to install JupyterLab:

```
pip install jupyterlab
```

5. Once the install has completed, exit Jupyter, stop your Gradient instance in the Paperspace console, and restart it.

6. When you get to the vanilla Jupyter **Files** view, update the **Uniform Resource Locator (URL)** to replace `tree` at the end of the URL with `lab`, and hit *Enter*. You should now see the JupyterLab view instead of the vanilla Jupyter view:

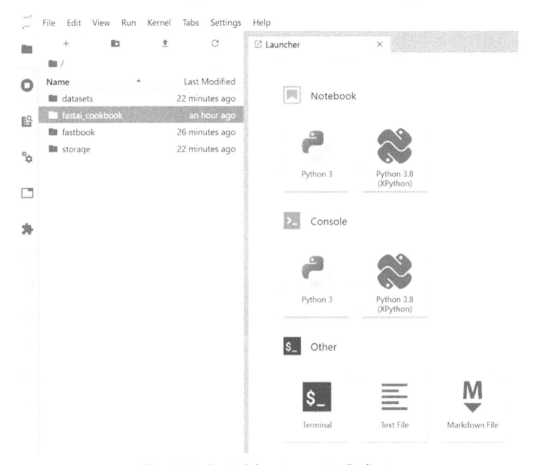

Figure 1.29 – JupyterLab environment in Gradient

Congratulations! You have set up Gradient so that you can use the JupyterLab view.

How it works...

You can go back and forth between the vanilla Jupyter view and JupyterLab any time you like by simply modifying the URL so that the end is `tree` (for Jupyter) or `lab` (for JupyterLab).

There's more...

If you want more details on the benefits of JupyterLab, this tutorial explains the features and how to use them: `https://dzone.com/articles/getting-started-with-jupyterlab`.

I mentioned earlier that one of the benefits of JupyterLab is that it supports a visual Python debugger you can use in notebooks. For more details on this debugger and how to set it up, see `https://medium.com/@cristiansaavedra/visual-jupyter-debugger-for-python-e96fdd4f6f68` and `https://blog.jupyter.org/a-visual-debugger-for-jupyter-914e61716559`.

"Hello world" for fastai – creating a model for MNIST

Now that you have set up your environment for fastai, it's time to run through an example. In this section, you will go through the process of creating a simple deep learning model trained on the MNIST dataset. This dataset consists of images of handwritten digits. The goal of the trained model is to predict the digit given an image. For example, we want the trained model to predict that the following digits are 6, 3, 9, and 6:

Figure 1.30 – Sample handwritten digits from the MNIST dataset

We won't be covering every detail of the fastai solution for MNIST in this section, but we will be running a complete example that demonstrates one of the key values of fastai—getting a powerful deep learning result with only a few lines of code. This example should also whet your appetite for the more advanced fastai examples that are coming in subsequent chapters.

Getting ready...

Ensure that you have followed the steps to set up fastai in Gradient and confirm that you can open the MNIST hello_world notebook (`mnist_hello_world.ipynb`) in the ch1 directory. If you choose to use Colab, ensure that you have selected **Runtime** | **Change runtime type** and have selected **GPU** as the hardware accelerator.

The dataset used in this section is the classic dataset of deep learning, MNIST (`http://yann.lecun.com/exdb/mnist/`). I gratefully acknowledge the opportunity to use this dataset to provide an initial illustration of the capabilities of fastai.

> **Dataset citation**
>
> Y. LeCun, L. Bottou, Y. Bengio and P. Haffner.(1998) Gradient-Based Learning Applied to Document Recognition (http://yann.lecun.com/exdb/publis/pdf/lecun-98.pdf). Proceedings of the IEEE, 86(11):2278-2324, November 1998

How to do it...

You will begin by running the notebook all the way through. By running all the cells in this notebook you will be executing code that trains an image classification deep learning model that predicts the class (that is, which digit from 0 to 9) a given image of a hand-written digit belongs to.

First, you will make the MNIST dataset, which consists of a set of images of handwritten digits organized into directories (one for each digit from 0 to 9), available to the Python code in the notebook. Next, you will define a `dataloaders` object that specifies the training subset of the dataset (that is, the images that will be used to train the model) and the validation subset of the dataset (that is, the images that will be used to assess the performance of the model as it is trained). Next, you will define the deep learning model itself using a pre-defined architecture (that is, an organization of layers that make up the model) made available by fastai.

Next, you will train the model, that is iteratively apply the training set to update the weights in the model to optimize the performance of the model for the specified metric (in the case of this model, accuracy). Next, you will examine batches of images in the training and validation sets. You will then look at images where the model does the worst job of classification. Finally, you will apply the trained deep learning model to example hand-written images to see whether the model predicts the correct digit for these images. In the following steps you will run the code in the entire notebook and then go through the cells in the notebook one-by-one to review what the code is doing:

1. Open the MNIST `hello_world` notebook `mnist_hello_world.ipynb` in the `ch1` directory.

2. Run the entire notebook by selecting the appropriate choice for your environment:

    a) **Cell|Run all** (Jupyter)

    b) **Run|Run all** (JupyterLab)

    c) **Runtime|Run all** (Colab)

3. Confirm that the notebook runs correctly to the end. You should see the following output from the last cell. Don't worry if you see a different digit output, as long as you get output with no errors in this cell:

```
# select a different image and apply the model to it to get a prediction
img = PILImage.create(img_files[5800])
img
```

```
learn.predict(img)
```

```
('1',
 TensorImage(1),
 TensorImage([0.0092, 0.9091, 0.0106, 0.0121, 0.0087, 0.0121, 0.0098, 0.0091, 0.0115,
0.0078]))
```

Figure 1.31 – Example MNIST digit prediction

Congratulations! You have just successfully trained your first deep learning model with fastai and used the trained model to predict the digits depicted in a set of handwritten digits from the MNIST dataset. Now, let's go through the notebook cell by cell to review what this example tells us about fastai:

1. The first code cell imports the libraries that the notebook needs:

    ```
    !pip install -Uqq fastbook
    import fastbook
    from fastbook import *
    from fastai.vision.all import *
    ```

2. The second cell calls the `fastai` function to prepare a notebook to run a fastai application. In Colab, for example, this function triggers the steps to mount Drive so that it's accessible within the notebook:

    ```
    fastbook.setup_book()
    ```

3. The third cell defines the location of the dataset that will be used to train the model. fastai provides a set of oven-ready datasets (including several varieties of the MNIST dataset) that you can ingest into your notebook with a single call using the `untar_data()` function. We'll dig into more details about these datasets in *Chapter 2, Exploring and Cleaning Up Data with fastai*:

    ```
    path = untar_data(URLs.MNIST)
    ```

4. The fifth cell is the heart of the notebook and demonstrates the power of fastai. Here are three lines of code that completely define and train a deep learning model:

a) The first line creates a `dataloaders` object from the `path` object created in the previous cell and identifies the subdirectories that contain the training and validation datasets. See the fastai documentation (`https://docs.fast.ai/vision.data.html#ImageDataLoaders`) for more details on `ImageDataLoaders`, the specific kind of `dataloaders` object used for image problems:

```
dls = ImageDataLoaders.from_folder(path,
train='training', valid='testing')
```

b) The second line defines the structure of the deep learning model, including its architecture (based on the famous **residual neural network** (**ResNet**) architecture - for more details see the documentation (`https://pytorch.org/vision/stable/models.html`), its loss function (in this case, a loss function that is appropriate for a multi-class classification problem), and the metric that will be optimized (in this case, validation accuracy):

```
learn = cnn_learner(dls, resnet18,
pretrained=False,
sloss_func=LabelSmoothingCrossEntropy(),
metrics=accuracy)
```

c) The third line trains the model, specifying that the training will be for 1 epoch (that is, one iteration through the entire training set) and the learning rate will be 0.1:

```
learn.fit_one_cycle(1, 0.1)
```

5. The output of this cell shows the results of the training process, including the training and validation loss, the validation accuracy, and the time taken to complete the training. Note that the accuracy is very high:

epoch	train_loss	valid_loss	accuracy	time
0	0.567086	0.524767	0.991400	00:50

Figure 1.32 – Output of training the MNIST model

6. The next two cells display examples of the training and validation **test** datasets:

```
# show a batch of training data
dls.train.show_batch(max_n=4, nrows=1)
```

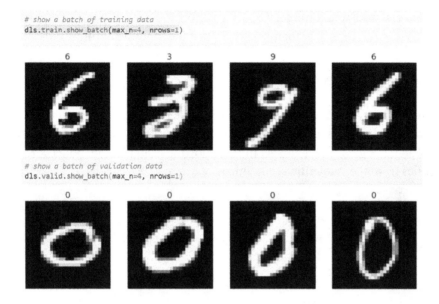

```
# show a batch of validation data
dls.valid.show_batch(max_n=4, nrows=1)
```

Figure 1.33 – Examples from the MNIST train and test datasets

7. The next cell shows examples of digits that the model got the most wrong. Note how these digits are not easy for us humans to identify, so it's not surprising that the model got them wrong:

```
# show the images with the highest loss
interp = ClassificationInterpretation.from_learner(learn)
interp.plot_top_losses(4, nrows=1)
```

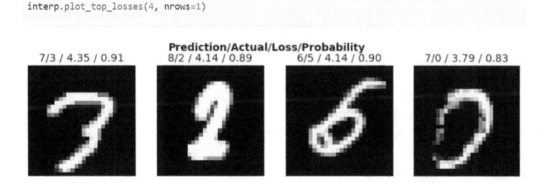

Figure 1.34 – Digits for which the MNIST model made the worst predictions

8. The next cell displays a summary of information about the model, including the layers that make it up, how many parameters it has, and the optimizer and loss function used:

```
learn.summary()
```

9. Finally, we have a set of cells at the end of the notebook that display digit images from the validation set, and then apply the trained model to get a prediction of which digit is shown in the image. In the following example, the model correctly identifies from the validation set an image of a zero as a zero:

```
# select an image from the test set
img = PILImage.create(img_files[0])
img
```

```
# apply the trained model to the image
learn.predict(img)
```

```
('0',
 TensorImage(0),
 TensorImage([0.9018, 0.0085, 0.0073, 0.0085, 0.0100, 0.0090, 0.0146, 0.0104, 0.0114, 0.0184]))
```

Figure 1.35 – Example prediction by the MNIST model

That's it—a complete, self-contained deep learning model that solves a famous computer vision problem (predicting a digit in a handwritten image) with remarkable accuracy. With fastai, you can accomplish this with just a few lines of code. This notebook contains some additional code to validate the model and investigate the dataset, but all you really need are the first five cells, which together contain barely 10 lines of code.

How it works...

You may be asking yourself about the details of the code that we just went through. How exactly does the data ingestion work? What's a dataloader? How does the model know to include all the layers that are shown by the summary() function? We'll be answering these questions in subsequent chapters. In *Chapter 2, Exploring and Cleaning Up Data with fastai*, we'll dig into the data ingestion story for fastai, and in subsequent chapters we'll take a detailed tour through fastai's solutions for a set of common deep learning applications, including tabular data, text data, recommender systems, and computer vision.

One of the beauties of fastai is that you can abstract away much of the complexity of deep learning (if you want), and get a working and useful model with just a few lines of code, as with the MNIST model we just saw. However, fastai doesn't keep the details hidden or limit your flexibility. In addition to providing an elegant way to create deep learning models with very little code, fastai incorporates a set of layers, each of which reveals more flexibility and detail. This means that as you learn more about fastai, you can continue to dig deeper and customize your solution to meet your exact needs.

There's more...

This section makes use of some standard machine learning terminology, including **loss function**, **optimizer**, **accuracy**, and **multi-class classification**. If you need a refresher on these and other fundamental machine learning concepts, I recommend the series of tutorials here: `https://machinelearningmastery.com/`. This site includes clear descriptions of the major concepts of machine learning, along with Python code samples that illustrate how to apply the concepts.

Understanding the world in four applications: tables, text, recommender systems, and images

In their seminal paper describing fastai, Howard and Gugger (`https://arxiv.org/pdf/2002.04688.pdf`) describe the four application areas that fastai supports *out of the box*. In this section, we will go through these four applications of deep learning that fastai directly supports: tabular data, text data, recommender systems, and computer vision. The MNIST example that you saw in the previous section is an example of a computer vision application. The MNIST example included the following:

- Curated dataset: MNIST. You can find an overall list of curated datasets here:

  `https://course.fast.ai/datasets`
- Easy ingestion of the curated dataset via `untar_data()`
- Image-specific handling of the dataset via a data loader object
- Definition of an image-specific model structure via a `Learner` object
- Utilities to examine the dataset

Similarly, fastai also provides components specifically aimed at the other three application areas: tabular data, text data, and recommender systems. In this section, we'll examine each of these application areas and learn about how fastai provides support for them.

Getting ready

Have the MNIST example that you went through in the last section open because we will be referring to is as we go through the four application areas. Before we get into the description of the four application areas, it's important to get some definitions:

- `DataLoader`: A structure that allows you to access batches of x (independent) and y (dependent) values. The x values are the data you use to train the model, and the y values are what you are trying to predict with the model.

- `DataLoaders`: A structure that contains training and validation `DataLoader` objects.

- `Learner`: An object that combines `DataLoaders`, architecture, and other characteristics (including loss function and optimizer) to define a model. To contrast `Learner` objects with models in Keras, `Learner` objects fully incorporate the data used to train the model, whereas with Keras models, the various facets of the dataset (such as training independent values, training dependent values, and much more) are arguments to the model that need to be addressed separately from the model itself.

How to do it...

Let's go through each of the four application areas and examine the support that fastai provides for them.

1. **Text data**, also called free-form text: fastai provides support for **natural language processing** (**NLP**):

   a) Curated datasets, including: AG_NEWS (~ 0.5 M categorized news articles), and DBPedia (training/testing samples from a knowledge base (`https://wiki.dbpedia.org/about`) containing structured content from Wikimedia projects), YELP_REVIEWS (~1.5 M Yelp reviews along with corresponding star scores)

   b) Text-specific `DataLoaders` object: `TextDataLoaders` `https://docs.fast.ai/text.data.html#TextDataLoaders`

   c) Text-specific learner object: `TextLearner` `https://docs.fast.ai/text.learner.html#TextLearner`

2. **Tabular data**, also called structured data, is data arranged in rows and columns, such as you would find in a **comma-separated values** (**CSV**) file or a database table. fastai provides custom support for deep learning with tabular data, including the following features:

   a) Tabular data-specific `DataLoaders` object: `TabularDataLoaders` `https://docs.fast.ai/tabular.data.html#TabularDataLoaders`

   b) Tabular data-specific `Learner` object: `TabularDataLearner` `https://docs.fast.ai/tabular.learner.html#TabularLearner`

   c) Utilities to examine the dataset: `TabularPandas` `https://docs.fast.ai/tabular.core.html#TabularPandas`

3. **Recommender systems**, also called collaborative filtering systems, combine aspects of text and tabular data as well as combining supervised and unsupervised learning to make predictions about a user's reaction to a given artifact.

   For example, recommender systems can be used to predict whether viewers will like movies they haven't seen yet or whether readers will like books they haven't read yet. fastai supports deep learning with recommender systems with a variety of features, including the following:

   a) Curated datasets, including `ML_SAMPLE` and `ML_100k` (rankings of thousands of movies by thousands of users)

   b) Recommender system-specific `DataLoaders` object: `CollabDataLoaders` `https://docs.fast.ai/collab.html#CollabDataLoaders`

   c) Recommender system-specific `Learner` object: `collab_learner` `https://docs.fast.ai/collab.html#Create-a-Learner`

4. **Image data**, also called computer vision applications. You have already seen fastai in action in this application area in the MNIST example. Of the four application areas, computer vision gets the most focus from fastai. The following list is just a subset of the many features in the fastai framework that make it easy to create deep learning solutions for image data problems:

   a) Curated datasets, including: MNIST (handwritten digits) and CARS (15,000+ images of cars, categorized into types) image classification datasets; `BIWI_HEAD_POSE` (images of people, along with descriptions of their positions), `PASCAL_2007`, and `PASCAL_2012` (images, along with corresponding segmentation maps for each image) image localization datasets

    b) Image-specific `DataLoaders` object: `ImageDataLoaders https://docs.fast.ai/vision.data.html#ImageDataLoaders`

    c) Image-specific `Learner` object: `cnn_learner https://docs.fast.ai/vision.learner.html#cnn_learner`

    d) Utilities to examine the dataset: many convenient functions that make it easy to render individual images and categories of images

How it works...

There is a lot to digest in this section, but don't worry. Each application-specific aspect of fastai gets its own dedicated chapter in which we'll cover the details of the application-specific features (datasets, `DataLoaders`, `Learners`, and others) and show you how to harness these features to create deep learning solutions for each application area. The important thing to note is that fastai provides these application-specific features to make it easy for you to create applications across all four of the areas: **tabular data**, **text data**, **recommender systems**, and **computer vision**.

Working with PyTorch tensors

Throughout most of this book, the focus will be on the features provided by the fastai framework. However, some of the solutions that we'll review also exploit general Python libraries (such as the `pandas` library for deep learning applications with tabular data) as well as aspects of PyTorch, the low-level deep learning framework upon which fastai is built. To give you a small taste of PyTorch, in this section we'll go through some basic examples of using tensors, the PyTorch structure for multidimensional matrices.

Getting ready

If you are already familiar with NumPy arrays, then you will have a good basis for examining PyTorch tensors because tensors play much the same role for PyTorch as NumPy arrays do for general-purpose Python applications. If you are not familiar with NumPy arrays or it's been a while since you have had a chance to use them, take a bit of time to review them—for example, by going through this tutorial: `https://numpy.org/doc/stable/user/quickstart.html`.

Once you have completed your NumPy array review, ensure that you have followed the steps to set up fastai in Gradient and confirm that you can open the PyTorch tensor walkthrough notebook (`pytorch_tensor_walkthrough.ipynb`) in the `ch1` directory. If you choose to use Colab, ensure that you have selected **Runtime** | **Change runtime type** and have selected **GPU** as the hardware accelerator.

How to do it...

In this section, we'll go through some basic operations with PyTorch tensors:

1. Open the `pytorch_tensor_walkthrough.ipynb` PyTorch tensor walkthrough notebook in the `ch1` directory.

2. Run the first four cells of the notebook to import the necessary libraries and define three tensors.

    a) Note that since this notebook only makes use of PyTorch and doesn't need any fastai libraries, we only need one `import` statement:

```
import torch
```

    b) Define a: a two-dimensional 5x7 tensor with value `1` in every position:

```
a = torch.ones(5, 7, dtype=torch.float)
a

tensor([[1., 1., 1., 1., 1., 1., 1.],
        [1., 1., 1., 1., 1., 1., 1.],
        [1., 1., 1., 1., 1., 1., 1.],
        [1., 1., 1., 1., 1., 1., 1.],
        [1., 1., 1., 1., 1., 1., 1.]])
```

Figure 1.36 – Defining a 5x7 tensor

    c) Define b: a two-dimensional 5x7 tensor with `0`s in every position except a diagonal of `1`s:

```
b = torch.eye(5,7)
b

tensor([[1., 0., 0., 0., 0., 0., 0.],
        [0., 1., 0., 0., 0., 0., 0.],
        [0., 0., 1., 0., 0., 0., 0.],
        [0., 0., 0., 1., 0., 0., 0.],
        [0., 0., 0., 0., 1., 0., 0.]])
```

Figure 1.37 – Defining a 5x7 tensor with 1s on the diagonal

d) Define c: a two-dimensional 5x5 identity tensor:

```
c = torch.eye(5,5)
c

tensor([[1., 0., 0., 0., 0.],
        [0., 1., 0., 0., 0.],
        [0., 0., 1., 0., 0.],
        [0., 0., 0., 1., 0.],
        [0., 0., 0., 0., 1.]])
```

Figure 1.38 – Defining a 5x5 identity tensor

3.   Now, run the cells in the **Examine tensor elements** section of the notebook to look at parts of one of the tensors.

a) Get the 0th row of tensor b:

```
b[0]

tensor([1., 0., 0., 0., 0., 0., 0.])
```

Figure 1.39 – The 0th row of tensor b

b) Get the 0th element of the 0th row:

```
b[0,0]

tensor(1.)
```

Figure 1.40 – Element [0,0] of tensor b

c) Get rows starting at row 2 to the end:

```
b[2:]

tensor([[0., 0., 1., 0., 0., 0., 0.],
        [0., 0., 0., 1., 0., 0., 0.],
        [0., 0., 0., 0., 1., 0., 0.]])
```

Figure 1.41 – Rows of tensor b from row to the end

4. Run the cells in the **Do operations on the tensors** section of the notebook to see how you can apply basic matrix arithmetic operations to tensors.

a) Add tensors a and b:

```
a_plus_b = a + b
a_plus_b

tensor([[2., 1., 1., 1., 1., 1., 1.],
        [1., 2., 1., 1., 1., 1., 1.],
        [1., 1., 2., 1., 1., 1., 1.],
        [1., 1., 1., 2., 1., 1., 1.],
        [1., 1., 1., 1., 2., 1., 1.]])
```

Figure 1.42 – Adding two tensors

b) Attempt to multiply tensors a and c—note that you get an error because the tensors do not have compatible dimensions. To multiply two two-dimensional tensors, the second dimension of the first tensor has to be identical to the first dimension of the second vector:

```
a_mult_c = a@c
a_mult_c
```

```
---------------------------------------------------------------------------
RuntimeError                                 Traceback (most recent call last)
<ipython-input-27-a25dfb8b9002> in <module>
      1 # multiply two tensors
----> 2 a_mult_c = a@c
      3 a_mult_c

RuntimeError: mat1 and mat2 shapes cannot be multiplied (5x7 and 5x5)
```

Figure 1.43 – Attempt to multiply two incompatible tensors generates an error

c) Define a 7x7 identity tensor:

```
d = torch.eye(7,7)
d

tensor([[1., 0., 0., 0., 0., 0., 0.],
        [0., 1., 0., 0., 0., 0., 0.],
        [0., 0., 1., 0., 0., 0., 0.],
        [0., 0., 0., 1., 0., 0., 0.],
        [0., 0., 0., 0., 1., 0., 0.],
        [0., 0., 0., 0., 0., 1., 0.],
        [0., 0., 0., 0., 0., 0., 1.]])
```

Figure 1.44 – Defining a 7x7 identity tensor

d) Now, multiply tensors a and d—this time, there is no error because the tensors' dimensions are compatible:

```
a_mult_d = a@d
a_mult_d

tensor([[1., 1., 1., 1., 1., 1., 1.],
        [1., 1., 1., 1., 1., 1., 1.],
        [1., 1., 1., 1., 1., 1., 1.],
        [1., 1., 1., 1., 1., 1., 1.],
        [1., 1., 1., 1., 1., 1., 1.]])
```

Figure 1.45 – Multiplying two compatible tensors

e) Create a new tensor that is the transpose of tensor a (that is, the columns of tensor a become the rows of the new tensor):

```
a_trans = torch.transpose(a,0,1)
a_trans

tensor([[1., 1., 1., 1., 1.],
        [1., 1., 1., 1., 1.],
        [1., 1., 1., 1., 1.],
        [1., 1., 1., 1., 1.],
        [1., 1., 1., 1., 1.],
        [1., 1., 1., 1., 1.],
        [1., 1., 1., 1., 1.]])
```

Figure 1.46 – Transposing a tensor

f) Multiply the transpose of tensor a with tensor c — while tensor a multiplied by tensor c caused an error, there will be no error this time because the tensors' dimensions are compatible:

```
a_trans_mult_c = a_trans @ c
a_trans_mult_c

tensor([[1., 1., 1., 1., 1.],
        [1., 1., 1., 1., 1.],
        [1., 1., 1., 1., 1.],
        [1., 1., 1., 1., 1.],
        [1., 1., 1., 1., 1.],
        [1., 1., 1., 1., 1.],
        [1., 1., 1., 1., 1.]])
```

Figure 1.47 – Multiplying two compatible tensors

Congratulations! You have had your first direct taste of PyTorch, the framework that underlies fastai.

How it works...

In this section you got a taste of tensors, one of the building blocks of PyTorch. If you are familiar with the relationship between Keras and TensorFlow, you can think of the relationship between fastai and PyTorch being similar. Similar to the way that Keras is a high-level **application programming interface (API)** for TensorFlow, fastai is built on top of PyTorch and abstracts away some of the complexity of PyTorch (for example, by making reasonable assumptions about defaults). With fastai, you can focus on creating deep learning applications without having to worry about all the details.

There's more...

If you are curious and want to get an overview of PyTorch now, you can check out this introductory tutorial: `https://pytorch.org/tutorials/beginner/nlp/pytorch_tutorial.html`.

Contrasting fastai with Keras

In this section, we'll cover some of the similarities and differences between fastai and Keras. While both frameworks provide high-level APIs for deep learning, there are some significant differences between them in terms of their architecture and approach to the problem, as well as differences between the communities using each. By contrasting these two frameworks, you will get a clearer idea of the strengths of fastai and be better prepared for the detailed examinations of fastai applications that are coming in subsequent chapters.

Getting ready

If you have used Keras recently, then you'll be in good shape to benefit from this section. If you haven't used Keras before, or it's been a while since you've used it, I recommend that you take a brief look at this tutorial so that you have a fresh overview of Keras: `https://keras.io/getting_started/intro_to_keras_for_engineers/`.

How to do it...

In this section, we will compare a Keras approach to the MNIST problem with the fastai MNIST solution that we reviewed earlier in this chapter. You can see the Keras approach in the ch1 directory of the repository, in `keras_sequential_api_hello_world.ipynb`.

Note that, by default, you will not be able to execute this Keras notebook in your fastai Gradient instance because the required TensorFlow and Keras libraries are not installed in that instance. You will be able to run the Keras MNIST notebook in Colab, if you have that set up.

1. Compare the library `import` statements. Both MNIST examples require a similar number of `import` statements:

    a) Keras:

    ```
    import tensorflow as tf
    import pydotplus
    from tensorflow.keras.utils import plot_model
    ```

    b) fastai:

    ```
    import fastbook
    from fastbook import *
    from fastai.vision.all import *
    ```

2. Compare the setup and definition of the dataset:

    a) Keras—The MNIST dataset is *oven ready* with Keras. Keras offers seven such datasets—for details, see `https://keras.io/api/datasets/`. By comparison, fastai has over 25 such datasets. For details, see `https://course.fast.ai/datasets`:

    ```
    mnist = tf.keras.datasets.mnist
    (x_train, y_train), (x_test, y_test) = mnist.load_data()
    x_train, x_test = x_train / 255.0, x_test / 255.0
    ```

    b) fastai—This requires a `setup` statement that Keras doesn't need (although this `setup` statement saves a step in the Drive mounting process when you are using Colab) but only requires two statements to define the dataset, versus three statements for Keras:

    ```
    fastbook.setup_book()
    path = untar_data(URLs.MNIST)
    dls = ImageDataLoaders.from_folder(path,
    train='training', valid='testing')
    ```

3. Compare the model definition statements:

a) Keras—Every layer in the model needs to be explicitly spelled out:

```
# define layers for the hello world model
hello_world_model = tf.keras.models.Sequential([
        tf.keras.layers.Flatten(input_shape=(28, 28)),
        tf.keras.layers.Dense(128, activation='relu'),
        tf.keras.layers.Dropout(0.15),
        tf.keras.layers.Dense(10)
])
# compile the hello world model, including specifying the
loss # function, optimizer, and metrics
hello_world_model.compile(optimizer='adam',
sloss=tf.keras.losses.SparseCategoricalCrossentropy(from
_logits=True), metrics=['accuracy'])
```

b) fastai—A single statement defines the model. The ability to specify the architecture (in this case, `resnet18`) with a single parameter streamlines the model definition. Note that the architecture specified for the fastai model is not identical to the architecture for the Keras model. For example, if you compared the layers listed in the output of the `learn.summary()` cell in this notebook with the layers specified in the definition of the Keras model, you can see that the fastai model has many more layers than the Keras model. In sum, the contrast between the fastai and Keras solutions for MNIST is not strictly *apples to apples*:

```
learn = cnn_learner(dls, resnet18, pretrained=False,
sloss_func=LabelSmoothingCrossEntropy(),
metrics=accuracy)
```

4. Compare the `fit` statements:

a) Keras:

```
history = hello_world_model.fit(x_train, y_train,
                                    batch_size=64,
                                    epochs=10,
                                    validation
_split=0.15)
```

b) fastai:

```
learn.fit_one_cycle(1, 0.1)
```

5. Compare the performance of the Keras model and the fastai model. Again, note that because of differences between the models (including the architecture and details of the fitting process), it's not possible to draw a general conclusion from the differences in performance between the two models:

a) Keras:

```
Loss for test dataset: 0.07588852692145155
Accuracy for test dataset: 0.9775
```

b) fastai:

epoch	train_loss	valid_loss	accuracy	time
0	0.567086	0.524767	0.991400	00:50

Figure 1.48 – Results of training the MNIST model in fastai

You have now seen a quick comparison of fastai and Keras for the same MNIST problem.

How it works...

What does this comparison between a Keras solution for MNIST and a fastai solution for MNIST tell us?

- Keras offers far fewer *oven-ready* datasets than fastai, and the fastai statements for defining such datasets are simpler. This is a critical benefit for fastai, particularly for beginners. It really helps in the process of learning about deep learning to have a wide variety of datasets that can be ingested easily. fastai really delivers on this count thanks to the big and varied set of *oven-ready* datasets available with fastai. We'll spend some time in the next chapter taking a closer look at these datasets.

- Other than the model definition, there isn't that much difference in the number of lines of code between Keras and fastai for each of the steps in the solution. This means that for the MNIST problem, Keras isn't far behind fastai's standard of delivering a complete solution with a handful of lines of code.

- The model definition is more complex in Keras, primarily because fastai lets us define the layers that make up the model with a single architecture parameter, whereas we have to explicitly define the layers in Keras. A mitigating factor for the complexity of the model definition in Keras is readability. In Keras, the layers are explicitly listed. By comparison, in the high-level fastai API, the layers are not listed.

- fastai offers better usability than Keras by making it possible for users to use the high-level fastai API without having to worry about all the explicit details.

- The statement for fitting the model is simpler in fastai. In addition, fastai incorporates best practices in default settings that often result in faster fitting times and better performance.

Keras benefits from greater transparency because the layers are explicitly listed. fastai has superior usability and out-of-the box performance thanks to carefully selected defaults for many settings. We are not going to do additional Keras versus fastai bakeoffs in this book, but I expect that, based on my experience using both Keras and fastai, fastai's benefits would stand out even more in complex applications. In addition, fastai has a big advantage because of its large set of curated, *oven-ready* datasets.

Test your knowledge

Now that you have completed the recipes in this chapter, you can follow the next steps to exercise that you have learned:

1. Make a copy of the `mnist_hello_world.ipynb` notebook—call it `mnist_hello_world_variations.ipynb`.

2. Update your new copy of the notebook to ingest a variation of the MNIST dataset, called `MNIST_SAMPLE`. Which statement will you need to update to ingest this dataset rather than the full-blown MNIST curated dataset?

3. Use the `path.ls()` statement to examine the directory structure of the `MNIST_SAMPLE` dataset. How is the output of this statement different from its output for the full-blown MNIST dataset?

4. Keeping in mind the difference in the directory structure of the `MNIST_SAMPLE` dataset, update the values of the `train` and `valid` parameters in the following statement so that it will work with this dataset:

```
dls = ImageDataLoaders.from_folder(path,
train='training', valid='testing')
```

5. Again keeping the directory structure in mind, update the following statement so that it will work with the `MNIST_SAMPLE` dataset:

```
img_files = get_image_files(path/"testing")
```

6. The `MNIST_SAMPLE` dataset is smaller than the full-blown MNIST dataset. Keeping this in mind, update the following statements so that they will work with the smaller dataset:

```
img = PILImage.create(img_files[7000])
img = PILImage.create(img_files[2030])
img = PILImage.create(img_files[5800])
```

7. Now that you have updated the notebook to work with the `MNIST_SAMPLE` dataset, run the whole notebook to confirm that it can run to the end with no errors.

Congratulations! If you have completed this section, then you have successfully adapted a recipe to work with another curated dataset.

2
Exploring and Cleaning Up Data with fastai

In the previous chapter, we got started with the fastai framework by setting up its coding environment, working through a concrete application example (MNIST), and investigating two frameworks with different relationships to fastai: PyTorch and Keras. In this chapter, we are going to dive deeper into an important aspect of fastai: **ingesting**, **exploring**, and **cleaning up data**. In particular, we are going to explore a selection of the datasets that are curated by fastai.

By the end of this chapter, you will be able to describe the complete set of curated datasets that fastai supports, use the facilities of fastai to examine these datasets, and clean up a dataset to eliminate missing and non-numeric values.

Here are the recipes that will be covered in this chapter:

- Getting the complete set of *oven-ready* fastai datasets
- Examining tabular datasets with fastai
- Examining text datasets with fastai

- Examining image datasets with fastai
- Cleaning up raw datasets with fastai

Technical requirements

Ensure that you have completed the setup sections in *Chapter 1, Getting Started with fastai*, and that you have a working Gradient instance or Colab setup. Ensure that you have cloned the repository for this book (`https://github.com/PacktPublishing/Deep-Learning-with-fastai-Cookbook`) and have access to the ch2 folder. This folder contains the code samples that will be described in this chapter.

Getting the complete set of oven-ready fastai datasets

In *Chapter 1, Getting Started with fastai*, you encountered the MNIST dataset and saw how easy it was to make this dataset available to train a fastai deep learning model. You were able to train the model without needing to worry about the location of the dataset or its structure (apart from the names of the folders containing the training and validation datasets). You were able to examine elements of the dataset conveniently.

In this section, we'll take a closer look at the complete set of datasets that fastai curates and explain how you can get additional information about these datasets.

Getting ready

Ensure you have followed the steps in *Chapter 1, Getting Started with fastai*, so that you have a fastai environment set up. Confirm that you can open the `fastai_dataset_walkthrough.ipynb` notebook in the ch2 directory of your cloned repository.

How to do it...

In this section, you will be running through the `fastai_dataset_walkthrough.ipynb` notebook, as well as the fastai dataset documentation, so that you understand the datasets that fastai curates. Once you have the notebook open in your fastai environment, complete the following steps:

1. Run the first three cells of the notebook to load the required libraries, set up the notebook for fastai, and define the MNIST dataset:

```python
# imports for notebook boilerplate
!pip install -Uqq fastbook
import fastbook
from fastbook import *
from fastai.vision.all import *
```

```python
# set up the notebook for fast.ai
fastbook.setup_book()
```

```python
# In Gradient, datasets get saved in /storage/data when untar_data is called
# if the dataset has not been copied there already
path = untar_data(URLs.MNIST)
```

Figure 2.1 – Cells to load the libraries, set up the notebook, and define the MNIST dataset

2. Consider the argument to `untar_data`: `URLs.MINST`. What is this? Let's try the `??` shortcut to examine the source code for a `URLs` object:

```python
??URLs
Init signature: URLs()
Source:
class URLs():
    "Global constants for dataset and model URLs."
    LOCAL_PATH = Path.cwd()
    MDL = 'http://files.fast.ai/models/'
    S3  = 'https://s3.amazonaws.com/fast-ai-'
    URL = f'{S3}sample/'

    S3_IMAGE    = f'{S3}imageclas/'
    S3_IMAGELOC = f'{S3}imagelocal/'
    S3_AUDI     = f'{S3}audio/'
    S3_NLP      = f'{S3}nlp/'
    S3_COCO     = f'{S3}coco/'
    S3_MODEL    = f'{S3}modelzoo/'

                    .
                    .
                    .

    # image classification datasets
    CALTECH_101  = f'{S3_IMAGE}caltech_101.tgz'
    CARS         = f'{S3_IMAGE}stanford-cars.tgz'
    CIFAR_100    = f'{S3_IMAGE}cifar100.tgz'
    CUB_200_2011 = f'{S3_IMAGE}CUB_200_2011.tgz'
    FLOWERS      = f'{S3_IMAGE}oxford-102-flowers.tgz'
    FOOD         = f'{S3_IMAGE}food-101.tgz'
    MNIST        = f'{S3_IMAGE}mnist_png.tgz'
```

Figure 2.2 – Source for URLs

3. By looking at the `image classification datasets` section of the source code for URLs, we can find the definition of URLs.MNIST:

```
MNIST              = f'{S3_IMAGE}mnist_png.tgz'
```

4. Working backward through the source code for the URLs class, we can get the whole URL for MNIST:

```
S3_IMAGE      = f'{S3}imageclas/'
S3  = 'https://s3.amazonaws.com/fast-ai-'
```

5. Putting it all together, we get the URL for URLs.MNIST:

```
https://s3.amazonaws.com/fast-ai-imageclas/mnist_png.tgz
```

6. You can download this file for yourself and untar it. You will see that the directory structure of the untarred package looks like this:

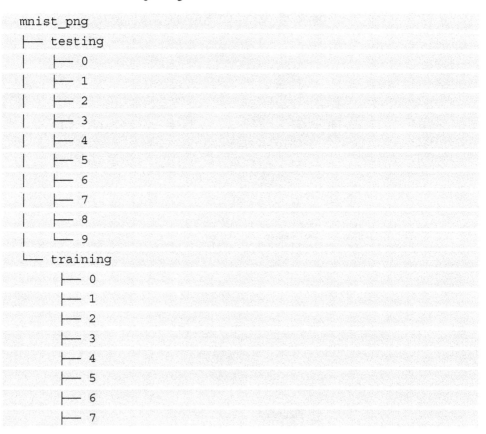

```
mnist_png
├── testing
│       ├── 0
│       ├── 1
│       ├── 2
│       ├── 3
│       ├── 4
│       ├── 5
│       ├── 6
│       ├── 7
│       ├── 8
│       └── 9
└── training
        ├── 0
        ├── 1
        ├── 2
        ├── 3
        ├── 4
        ├── 5
        ├── 6
        ├── 7
```

```
├── 8
└── 9
```

7. In the untarred directory structure, each of the testing and training directories contain subdirectories for each digit. These digit directories contain image files for that digit. This means that the label of the dataset – the value that we want the model to predict – is encoded in the directory that the image file resides in.

8. Is there a way to get the directory structure of one of the curated datasets without having to determine its URL from the definition of URLs, download the dataset, and unpack it? There is – using path.ls():

```
path.ls()
```

```
(#2) [Path('/storage/data/mnist_png/training'),Path('/storage/data/mnis
t_png/testing')]
```

Figure 2.3 – Using path.ls() to get the dataset's directory structure

9. This tells us that there are two subdirectories in the dataset: training and testing. You can call ls() to get the structure of the training subdirectory:

```
# get the structure of one of the training subdirectory
(path/'training').ls()
```

```
(#10) [Path('/storage/data/mnist_png/training/0'),Path('/storage/data/m
nist_png/training/2'),Path('/storage/data/mnist_png/training/9'),Path
('/storage/data/mnist_png/training/8'),Path('/storage/data/mnist_png/tr
aining/7'),Path('/storage/data/mnist_png/training/1'),Path('/storage/da
ta/mnist_png/training/5'),Path('/storage/data/mnist_png/training/4'),Pa
th('/storage/data/mnist_png/training/6'),Path('/storage/data/mnist_png/
training/3')]
```

Figure 2.4 – The structure of the training subdirectory

10. Now that we have learned how to get the directory structure of the MNIST dataset using the ls() function, what else can we learn from the output of ??URLs?

11. First, let's look at the other datasets listed in the output of ??URLs by group. First, let's look at the datasets listed under main datasets. This list includes tabular datasets (ADULT_SAMPLE), text datasets (IMDB_SAMPLE), recommender system datasets (ML_SAMPLE), and a variety of image datasets (CIFAR, IMAGENETTE, COCO_SAMPLE):

ADULT_SAMPLE	= f'{URL}adult_sample.tgz'
BIWI_SAMPLE	= f'{URL}biwi_sample.tgz'

```
        CIFAR                    = f'{URL}cifar10.tgz'
        COCO_SAMPLE              = f'{S3_COCO}coco_sample.tgz'
        COCO_TINY                = f'{S3_COCO}coco_tiny.tgz'
        HUMAN_NUMBERS            = f'{URL}human_numbers.tgz'
        IMDB                      = f'{S3_NLP}imdb.tgz'
        IMDB_SAMPLE              = f'{URL}imdb_sample.tgz'
        ML_SAMPLE                = f'{URL}movie_lens_sample.
tgz'
        ML_100k                  = 'http://files.grouplens.
org/datasets/movielens/ml-100k.zip'
        MNIST_SAMPLE             = f'{URL}mnist_sample.tgz'
        MNIST_TINY               = f'{URL}mnist_tiny.tgz'
        MNIST_VAR_SIZE_TINY = f'{S3_IMAGE}mnist_var_size_
tiny.tgz'
        PLANET_SAMPLE            = f'{URL}planet_sample.tgz'
        PLANET_TINY              = f'{URL}planet_tiny.tgz'
        IMAGENETTE               = f'{S3_IMAGE}imagenette2.
tgz'
        IMAGENETTE_160           = f'{S3_IMAGE}imagenette2-160.
tgz'
        IMAGENETTE_320           = f'{S3_IMAGE}imagenette2-320.
tgz'
        IMAGEWOOF                = f'{S3_IMAGE}imagewoof2.
tgz'
        IMAGEWOOF_160            = f'{S3_IMAGE}imagewoof2-160.
tgz'
        IMAGEWOOF_320            = f'{S3_IMAGE}imagewoof2-320.
tgz'
        IMAGEWANG                = f'{S3_IMAGE}imagewang.tgz'
        IMAGEWANG_160            = f'{S3_IMAGE}imagewang-160.
tgz'
        IMAGEWANG_320            = f'{S3_IMAGE}imagewang-320.
tgz'
```

12. Next, let's look at the datasets in the other categories: image classification datasets, NLP datasets, image localization datasets, audio classification datasets, and medical image classification datasets. Note that the list of curated datasets includes datasets that aren't directly associated with any of the four main application areas supported by fastai. The audio datasets, for example, apply to a use case outside the four main application areas:

```
# image classification datasets
CALTECH_101  = f'{S3_IMAGE}caltech_101.tgz'
CARS         = f'{S3_IMAGE}stanford-cars.tgz'
CIFAR_100    = f'{S3_IMAGE}cifar100.tgz'
CUB_200_2011 = f'{S3_IMAGE}CUB_200_2011.tgz'

FLOWERS      = f'{S3_IMAGE}oxford-102-flowers.tgz'
FOOD         = f'{S3_IMAGE}food-101.tgz'
MNIST        = f'{S3_IMAGE}mnist_png.tgz'
PETS         = f'{S3_IMAGE}oxford-iiit-pet.tgz'


# NLP datasets
AG_NEWS                 = f'{S3_NLP}ag_news_
csv.tgz'
AMAZON_REVIEWS          = f'{S3_NLP}amazon_
review_full_csv.tgz'
AMAZON_REVIEWS_POLARITY = f'{S3_NLP}amazon_review_
polarity_csv.tgz'
DBPEDIA                 = f'{S3_NLP}dbpedia_
csv.tgz'
MT_ENG_FRA              = f'{S3_NLP}giga-fren.
tgz'
SOGOU_NEWS              = f'{S3_NLP}sogou_
news_csv.tgz'
WIKITEXT                = f'{S3_NLP}
wikitext-103.tgz'
WIKITEXT_TINY           = f'{S3_NLP}wikitext-2.
tgz'
YAHOO_ANSWERS           = f'{S3_NLP}yahoo_
answers_csv.tgz'
YELP_REVIEWS            = f'{S3_NLP}yelp_
review_full_csv.tgz'
YELP_REVIEWS_POLARITY   = f'{S3_NLP}yelp_review_
polarity_csv.tgz'
```

```
      # Image localization datasets
      BIWI_HEAD_POSE      = f"{S3_IMAGELOC}biwi_head_pose.
tgz"
      CAMVID                = f'{S3_IMAGELOC}camvid.tgz'
      CAMVID_TINY           = f'{URL}camvid_tiny.tgz'
      LSUN_BEDROOMS         = f'{S3_IMAGE}bedroom.tgz'
      PASCAL_2007           = f'{S3_IMAGELOC}pascal_2007.
tgz'
      PASCAL_2012           = f'{S3_IMAGELOC}pascal_2012.
tgz'


      # Audio classification datasets
      MACAQUES              = 'https://storage.
googleapis.com/ml-animal-sounds-datasets/macaques.zip'
      ZEBRA_FINCH           = 'https://storage.googleapis.
com/ml-animal-sounds-datasets/zebra_finch.zip'


      # Medical Imaging datasets
      SIIM_SMALL            = f'{S3_IMAGELOC}siim_small.
tgz'
```

13. Now that we have listed all the datasets defined in URLs, how can we find out more information about them?

   a) The fastai documentation (https://course.fast.ai/datasets) documents some of the datasets listed in URLs. Note that this documentation is not consistent with what's listed in the source of URLs. For example, the naming of the datasets is not consistent and the documentation page does not cover all the datasets. When in doubt, treat the source of URLs as your single source of truth about fastai curated datasets.

   b) Use the path.ls() function to examine the directory structure, as shown in the following example, which lists the directories under the training subdirectory of the MNIST dataset:

```
(path/'training').ls()
```

```
(#10) [Path('/storage/data/mnist_png/training/0'),Path('/storage/data/mnist_png/training/2'),Path
('/storage/data/mnist_png/training/9'),Path('/storage/data/mnist_png/training/8'),Path('/storage/da
ta/mnist_png/training/7'),Path('/storage/data/mnist_png/training/1'),Path('/storage/data/mnist_png/
training/5'),Path('/storage/data/mnist_png/training/4'),Path('/storage/data/mnist_png/training/6'),
Path('/storage/data/mnist_png/training/3')]
```

Figure 2.5 – Structure of the training subdirectory

c) Check out the file structure that gets installed when you run untar_data. For example, in Gradient, the datasets get installed in storage/data, so you can go into that directory in Gradient to inspect the directories for the curated dataset you're interested in.

d) For example, let's say untar_data is run with URLs.PETS as the argument:

```
path = untar_data(URLs.PETS)
```

e) Here, you can find the dataset in storage/data/oxford-iiit-pet, and you can see the directory's structure:

```
oxford-iiit-pet
├── annotations
│     ├── trimaps
│     └── xmls
└── images
```

14. If you want to see the definition of a function in a notebook, you can run a cell with ??, followed by the name of the function. For example, to see the definition of the ls() function, you can use ??Path.ls:

```
?Path.ls
Signature: Path.ls(self: pathlib.Path, n_max=None, file_type=None, file_exts=None)
Source:
@patch
def ls(self:Path, n_max=None, file_type=None, file_exts=None):
    "Contents of path as a list"
    extns=L(file_exts)
    if file_type: extns += L(k for k,v in mimetypes.types_map.items() if v.startswith(fil
e_type+'/'))
    has_extns = len(extns)==0
    res = (o for o in self.iterdir() if has_extns or o.suffix in extns)
    if n_max is not None: res = itertools.islice(res, n_max)
    return L(res)
File:      /opt/conda/envs/fastai/lib/python3.8/site-packages/fastcore/xtras.py
Type:      function
```

Figure 2.6 – Source for Path.ls()

15. To see the documentation for any function, you can use the `doc()` function. For example, the output of `doc(Path.ls)` shows the signature of the function, along with links to the source code (`https://github.com/fastai/fastcore/blob/master/fastcore/xtras.py#L111`) and the documentation (`https://fastcore.fast.ai/xtras#Path.ls`) for this function:

```
doc(Path.ls)
```

Path.ls   [source]

   Path.ls (**n_max** = *None* , **file_type** = *None* , **file_exts** = *None*)

Contents of path as a list

Show in docs

Figure 2.7 – Output of doc(Path.ls)

You have now explored the list of oven-ready datasets curated by fastai. You have also learned how to get the directory structure of these datasets, as well as how to examine the source and documentation of a function from within a notebook.

How it works...

As you saw in this section, fastai defines URLs for each of the curated datasets in the URLs class. When you call `untar_data` with one of the curated datasets as the argument, if the files for the dataset have not already been copied, these files get downloaded to your filesystem (`storage/data` in a Gradient instance). The object you get back from `untar_data` allows you to examine the directory structure of the dataset, and then pass it along to the next stage in the process of creating a fastai deep learning model. By wrapping a large sampling of interesting datasets in such a convenient way, fastai makes it easy for you to create deep learning models with these datasets, and also lets you focus your efforts on creating and improving the deep learning model rather than fiddling with the details of ingesting the datasets.

There's more...

You might be asking yourself why we went to the trouble of examining the source code for the `URLs` class to get details about the curated datasets. After all, these datasets are documented in `https://course.fast.ai/datasets`. The problem is that this documentation page doesn't give a complete list of all the curated datasets, and it doesn't clearly explain what you need to know to make the correct `untar_data` calls for a particular curated dataset. The incomplete documentation for the curated datasets demonstrates one of the weaknesses of fastai – *inconsistent documentation*. Sometimes, the documentation is complete, but sometimes, it is lacking details, so you will need to look at the source code directly to figure out what's going on, like we had to do in this section for the curated datasets. This problem is compounded by Google search returning hits for documentation for earlier versions of fastai. If you are searching for some details about fastai, avoid hits for fastai version 1 (`https://fastai1.fast.ai/`) and keep to the documentation for the current version of fastai: `https://docs.fast.ai/`.

Examining tabular datasets with fastai

In the previous section, we looked at the whole set of datasets curated by fastai. In this section, we are going to dig into a tabular dataset from the curated list. We will ingest the dataset, look at some example records, and then explore characteristics of the dataset, including the number of records and the number of unique values in each column.

Getting ready

Ensure you have followed the steps in *Chapter 1, Getting Started with fastai*, to get a fastai environment set up. Confirm that you can open the `examining_tabular_datasets.ipynb` notebook in the `ch2` directory of your repository.

I am grateful for the opportunity to include the ADULT_SAMPLE dataset featured in this section.

> **Dataset citation**
>
> Ron Kohavi. (1996) *Scaling Up the Accuracy of Naive-Bayes Classifiers: a Decision-Tree Hybrid* (`http://robotics.stanford.edu/~ronnyk/nbtree.pdf`).

How to do it...

In this section, you will be running through the `examining_tabular_datasets.ipynb` notebook to examine the `ADULT_SAMPLE` dataset.

Once you have the notebook open in your fastai environment, complete the following steps:

1. Run the first two cells to import the necessary libraries and set up the notebook for fastai.

2. Run the following cell to copy the dataset into your filesystem (if it's not already there) and to define the path for the dataset:

    ```
    path = untar_data(URLs.ADULT_SAMPLE)
    ```

3. Run the following cell to get the output of `path.ls()` so that you can examine the directory structure of the dataset:

```
path.ls()
```

```
(#3) [Path('/storage/data/adult_sample/adult.csv'),Path('/storage/data/adult_sample/export.pk
l'),Path('/storage/data/adult_sample/models')]
```

Figure 2.8 – Output of path.ls()

4. The dataset is in the `adult.csv` file. Run the following cell to ingest this CSV file into a pandas DataFrame:

    ```
    df = pd.read_csv(path/'adult.csv')
    ```

5. Run the `head()` command to get a sample of records from the beginning of the dataset:

df.head()

	age	workclass	fnlwgt	education	education-num	marital-status	occupation	relationship	race	sex	capital-gain	capital-loss	hours-per-week	native-country	
0	49	Private	101320	Assoc-acdm	12.0	Married-civ-spouse	NaN	Wife	White	Female	0	1902	40	United-States	>=
1	44	Private	236746	Masters	14.0	Divorced	Exec-managerial	Not-in-family	White	Male	10520	0	45	United-States	>=
2	38	Private	96185	HS-grad	NaN	Divorced	NaN	Unmarried	Black	Female	0	0	32	United-States	<
3	38	Self-emp-inc	112847	Prof-school	15.0	Married-civ-spouse	Prof-specialty	Husband	Asian-Pac-Islander	Male	0	0	40	United-States	>=
4	42	Self-emp-not-inc	82297	7th-8th	NaN	Married-civ-spouse	Other-service	Wife	Black	Female	0	0	50	United-States	<

Figure 2.9 – Sample of records from the beginning of the dataset

6. Run the following command to get the number of records (rows) and fields (columns) in the dataset:

```
df.shape
```

7. Run the following command to get the number of unique values in each column of the dataset. Can you tell from the output which columns are categorical?

```
df.nunique()
```

8. Run the following command to get the count of missing values in each column of the dataset. Which columns have missing values?

```
df.isnull().sum()
```

9. Run the following command to display some sample records from the subset of the dataset for people whose age is less than or equal to 40:

```
df_young = df[df.age <= 40]
df_young.head()
```

Congratulations! You have ingested a tabular dataset curated by fastai and done a basic examination of the dataset.

How it works...

The dataset that you explored in this section, ADULT_SAMPLE, is one of the datasets you would have seen in the source for URLs in the previous section. Note that while the source for URLs identifies which datasets are related to image or NLP (text) applications, it does not explicitly identify the tabular or recommender system datasets. ADULT_SAMPLE is one of the datasets listed under main datasets:

```
# main datasets
ADULT_SAMPLE          = f'{URL}adult_sample.tgz'
BIWI_SAMPLE           = f'{URL}biwi_sample.tgz'
CIFAR                 = f'{URL}cifar10.tgz'
COCO_SAMPLE           = f'{S3_COCO}coco_sample.tgz'
COCO_TINY             = f'{S3_COCO}coco_tiny.tgz'
HUMAN_NUMBERS         = f'{URL}human_numbers.tgz'
IMDB                  = f'{S3_NLP}imdb.tgz'
IMDB_SAMPLE           = f'{URL}imdb_sample.tgz'
ML_SAMPLE             = f'{URL}movie_lens_sample.tgz'
ML_100k               = 'http://files.grouplens.org/datasets/movielens/ml-100k.zip'
MNIST_SAMPLE          = f'{URL}mnist_sample.tgz'
MNIST_TINY            = f'{URL}mnist_tiny.tgz'
MNIST_VAR_SIZE_TINY   = f'{S3_IMAGE}mnist_var_size_tiny.tgz'
PLANET_SAMPLE         = f'{URL}planet_sample.tgz'
PLANET_TINY           = f'{URL}planet_tiny.tgz'
IMAGENETTE            = f'{S3_IMAGE}imagenette2.tgz'
IMAGENETTE_160        = f'{S3_IMAGE}imagenette2-160.tgz'
IMAGENETTE_320        = f'{S3_IMAGE}imagenette2-320.tgz'
IMAGEWOOF             = f'{S3_IMAGE}imagewoof2.tgz'
IMAGEWOOF_160         = f'{S3_IMAGE}imagewoof2-160.tgz'
IMAGEWOOF_320         = f'{S3_IMAGE}imagewoof2-320.tgz'
IMAGEWANG             = f'{S3_IMAGE}imagewang.tgz'
IMAGEWANG_160         = f'{S3_IMAGE}imagewang-160.tgz'
IMAGEWANG_320         = f'{S3_IMAGE}imagewang-320.tgz'
```

Figure 2.10 – Main datasets from the source for URLs

How did I determine that ADULT_SAMPLE was a tabular dataset? First, the paper by Howard and Gugger (https://arxiv.org/pdf/2002.04688.pdf) identifies ADULT_SAMPLE as a tabular dataset. Second, I just had to ingest it and try it out to confirm it could be ingested into a pandas DataFrame.

There's more...

What about the other curated datasets that aren't explicitly categorized in the source for URLs? Here's a summary of the datasets listed in the source for URLs under `main datasets`:

- Tabular:

  a) `ADULT_SAMPLE`

- NLP (text):

  a) `HUMAN_NUMBERS`

  b) `IMDB`

  c) `IMDB_SAMPLE`

- Collaborative filtering:

  a) `ML_SAMPLE`

  b) `ML_100k`

- Image data:

  a) All of the other datasets listed in URLs under `main datasets`.

Examining text datasets with fastai

In the previous section, we looked at how a curated tabular dataset could be ingested. In this section, we are going to dig into a text dataset from the curated list.

Getting ready

Ensure you have followed the steps in *Chapter 1, Getting Started with fastai,* to get a fastai environment set up. Confirm that you can open the `examining_text_datasets.ipynb` notebook in the `ch2` directory of your repository.

I am grateful for the opportunity to use the WIKITEXT_TINY dataset (`https://blog.einstein.ai/the-wikitext-long-term-dependency-language-modeling-dataset/`) featured in this section.

> **Dataset citation**
>
> Stephen Merity, Caiming Xiong, James Bradbury, Richard Socher. (2016). *Pointer Sentinel Mixture Models* (`https://arxiv.org/pdf/1609.07843.pdf`).

How to do it...

In this section, you will be running through the `examining_text_datasets.ipynb` notebook to examine the `WIKITEXT_TINY` dataset. As its name suggests, this is a small set of text that's been gleaned from good and featured Wikipedia articles.

Once you have the notebook open in your fastai environment, complete the following steps:

1. Run the first two cells to import the necessary libraries and set up the notebook for fastai.

2. Run the following cell to copy the dataset into your filesystem (if it's not already there) and to define the path for the dataset:

    ```
    path = untar_data(URLs.WIKITEXT_TINY)
    ```

3. Run the following cell to get the output of `path.ls()` so that you can examine the directory structure of the dataset:

    ```
    path.ls()
    ```

    ```
    (#2) [Path('/storage/data/wikitext-2/train.csv'),Path('/storage/data/wi
    kitext-2/test.csv')]
    ```

    Figure 2.11 – Output of path.ls()

4. There are two CSV files that make up this dataset. Let's ingest each of them into a pandas DataFrame, starting with `train.csv`:

    ```
    df_train = pd.read_csv(path/'train.csv')
    ```

5. When you use `head()` to check the DataFrame, you'll notice that something's wrong – the CSV file has no header with column names, but by default, `read_csv` assumes the first row is the header, so the first row gets misinterpreted as a header. As shown in the following screenshot, the first row of output is in bold, which indicates that the first row is being interpreted as a header, even though it contains a regular data row:

```
df_train.head(2)
```

> \n = 2013 – 14 York City F.C. season = \n \n The 2013 – 14 season was the <unk>
> season of competitive association football and 77th season in the Football League
> played by York City Football Club , a professional football club based in York ,
> North Yorkshire , England . Their 17th @-@ place finish in 2012 – 13 meant it was
> their second consecutive season in League Two . The season ran from 1 July 2013
> to 30 June 2014 . \n Nigel Worthington , starting his first full season as York
> manager , made eight permanent summer signings . By the turn of the year York
> were only above the relegation zone on goal difference , before a 17 @-@ match

Figure 2.12 – First record in df_train

6. To fix this problem, rerun the read_csv function, but this time with the
    header=None parameter, to specify that the CSV file doesn't have a header:

```
df_train = pd.read_csv(path/'train.csv',header=None)
```

7. Check head() again to confirm that the problem has been resolved:

```
df_train.head(2)
```

0

0   \n = 2013 – 14 York City F.C. season = \n \n The 2013 – 14 season was the <unk>
     season of competitive association football and 77th season in the Football League
     played by York City Football Club , a professional football club based in York , North
     Yorkshire , England . Their 17th @-@ place finish in 2012 – 13 meant it was their second
     consecutive season in League Two . The season ran from 1 July 2013 to 30 June 2014 . \n
     Nigel Worthington , starting his first full season as York manager , made eight
     permanent summer signings . By the turn of the year York were only above the

Figure 2.13 – Revising the first record in df_train

8. Ingest test.csv into a DataFrame using the header=None parameter:

```
df_test = pd.read_csv(path/'test.csv',header=None)
```

9. We want to tokenize the dataset and transform it into a list of words. Since we want
    a common set of tokens for the entire dataset, we will begin by combining the test
    and train DataFrames:

```
df_combined = pd.concat([df_train,df_test])
```

10. Confirm the shape of the train, test, and combined dataframes – the number of rows in the combined DataFrame should be the sum of the number of rows in the train and test DataFrames:

```
print("df_train: ",df_train.shape)
print("df_test: ",df_test.shape)
print("df_combined: ",df_combined.shape)
```

11. Now, we're ready to tokenize the DataFrame. The `tokenize_df()` function takes the list of columns containing the text we want to tokenize as a parameter. Since the columns of the DataFrame are not labeled, we need to refer to the column we want to tokenize using its position rather than its name:

```
df_tok, count = tokenize_df(df_combined, [df_combined.
columns[0]])
```

12. Check the contents of the first few records of `df_tok`, which is the new DataFrame containing the tokenized contents of the combined DataFrame:

`df_tok.head(3)`

	text	text_length
0	[xxbos, =, 2013, –, 14, xxmaj, york, xxmaj, city, xxup, f.c, ., season, =, \n__\n__, xxmaj, the, 2013, –, 14, season, was, the, xxunk, season, of, competitive, association, football, and, 77th, season, in, the, xxmaj, football, xxmaj, league, played, by, xxmaj, york, xxmaj, city, xxmaj, football, xxmaj, club, ,, a, professional, football, club, based, in, xxmaj, york, ,, xxmaj, north, xxmaj, yorkshire, ,, xxmaj, england, ., xxmaj, their, 17th, -, place, finish, in, 2012, –, 13, meant, it, was, their, second, consecutive, season, in, xxmaj, league, xxmaj, two, ., xxmaj, the, season, ran, from...	4405
1	[xxbos, =, xxmaj, big, xxmaj, boy, (, song,), =, \n__\n__, ", xxmaj, big, xxmaj, boy, ", xxunk, ", i, ', m, a, xxmaj, big, xxmaj, boy, xxmaj, now, ", was, the, first, single, ever, recorded, by, the, xxmaj, jackson, 5, ,, which, was, released, by, xxmaj, steeltown, xxmaj, records, in, xxmaj, january, 1968, ., xxmaj, the, group, played, instruments, on, many, of, their, xxmaj, steeltown, compositions, ,, including, ", xxmaj, big, xxmaj, boy, ", ., xxmaj, the, song, was, neither, a, critical, nor, commercial, success, ,, but, the, xxmaj, jackson, family, were, delighted, with, the, outcome, n...	976

Figure 2.14 – The first few records of df_tok

13. Check the count for a few sample words to ensure they are roughly what you expected. Pick a very common word, a moderately common word, and a rare word:

```
print("very common word (count['the']):", count['the'])
print("moderately common word (count['prepared']):",
count['prepared'])
print("rare word (count['gaga']):", count['gaga'])
```

Congratulations! You have successfully ingested, explored, and tokenized a curated text dataset.

How it works...

The dataset that you explored in this section, WIKITEXT_TINY, is one of the datasets you would have seen in the source for URLs in the *Getting the complete set of oven-ready fastai datasets* section. Here, you can see that WIKITEXT_TINY is in the NLP datasets section of the source for URLs:

```
# NLP datasets
AG_NEWS                     = f'{S3_NLP}ag_news_csv.tgz'
AMAZON_REVIEWS              = f'{S3_NLP}amazon_review_full_csv.tgz'
AMAZON_REVIEWS_POLARITY = f'{S3_NLP}amazon_review_polarity_csv.tgz'
DBPEDIA                     = f'{S3_NLP}dbpedia_csv.tgz'
MT_ENG_FRA                  = f'{S3_NLP}giga-fren.tgz'
SOGOU_NEWS                  = f'{S3_NLP}sogou_news_csv.tgz'
WIKITEXT                    = f'{S3_NLP}wikitext-103.tgz'
WIKITEXT_TINY               = f'{S3_NLP}wikitext-2.tgz'
YAHOO_ANSWERS               = f'{S3_NLP}yahoo_answers_csv.tgz'
YELP_REVIEWS                = f'{S3_NLP}yelp_review_full_csv.tgz'
YELP_REVIEWS_POLARITY  = f'{S3_NLP}yelp_review_polarity_csv.tgz'
```

Figure 2.15 – WIKITEXT_TINY in the NLP datasets list in the source for URLs

Examining image datasets with fastai

In the past two sections, we examined tabular and text datasets and got a taste of the facilities that fastai provides for accessing and exploring these datasets. In this section, we are going to look at image data. We are going to look at two datasets: the FLOWERS image classification dataset and the BIWI_HEAD_POSE image localization dataset.

Getting ready

Ensure you have followed the steps in *Chapter 1, Getting Started with fastai*, to get a fastai environment set up. Confirm that you can open the examining_image_datasets. ipynb notebook in the ch2 directory of your repository.

I am grateful for the opportunity to use the FLOWERS dataset featured in this section.

Dataset citation

Maria-Elena Nilsback, Andrew Zisserman. (2008). *Automated flower classification over a large number of classes* (`https://www.robots.ox.ac.uk/~vgg/publications/papers/nilsback08.pdf`).

I am grateful for the opportunity to use the BIWI_HEAD_POSE dataset featured in this section.

Dataset citation

Gabriele Fanelli, Thibaut Weise, Juergen Gall, Luc Van Gool. (2011). *Real Time Head Pose Estimation from Consumer Depth Cameras* (`https://link.springer.com/chapter/10.1007/978-3-642-23123-0_11`). Lecture Notes in Computer Science, vol 6835. Springer, Berlin, Heidelberg `https://doi.org/10.1007/978-3-642-23123-0_11`.

How to do it...

In this section, you will be running through the `examining_image_datasets.ipynb` notebook to examine the FLOWERS and BIWI_HEAD_POSE datasets.

Once you have the notebook open in your fastai environment, complete the following steps:

1. Run the first two cells to import the necessary libraries and set up the notebook for fastai.

2. Run the following cell to copy the FLOWERS dataset into your filesystem (if it's not already there) and to define the path for the dataset:

   ```
   path = untar_data(URLs.FLOWERS)
   ```

3. Run the following cell to get the output of `path.ls()` so that you can examine the directory structure of the dataset:

```
path.ls()
```

```
(#3) [Path('/storage/data/oxford-102-flowers/valid.txt'),Path('/storage/data/oxford-102-flowers/jp
g'),Path('/storage/data/oxford-102-flowers/test.txt')]
```

Figure 2.16 – Output of path.ls()

4. Look at the contents of the valid.txt file. This indicates that train.txt, valid.txt, and test.txt contain lists of the image files that belong to each of these datasets:

```
df_valid = pd.read_csv(path/'valid.txt', header=None)
df_valid.head()
```

	0
0	jpg/image_04467.jpg 89
1	jpg/image_07129.jpg 44
2	jpg/image_05166.jpg 4
3	jpg/image_07002.jpg 34
4	jpg/image_02007.jpg 79

Figure 2.17 – The first few records of valid.txt

5. Examine the jgp subdirectory:

```
(path/'jpg').ls()
```

6. Take a look at one of the image files. Note that the get_image_files() function doesn't need to be pointed to a particular subdirectory – it recursively collects all the image files in a directory and its subdirectories:

```
img_files = get_image_files(path)
img = PILImage.create(img_files[100])
img
```

7. You should have noticed that the image displayed in the previous step was the native size of the image, which makes it rather big for the notebook. To get the image at a more appropriate size, apply the `to_thumb` function with the image dimension specified as an argument. Note that you might see a different image when you run this cell:

```
img.to_thumb(180)
```

Figure 2.18 – Applying to_thumb to an image

8. Now, ingest the `BIWI_HEAD_POSE` dataset:

```
path = untar_data(URLs.BIWI_HEAD_POSE)
```

9. Examine the path for this dataset:

```
path.ls()
```

10. Examine the `05` subdirectory:

```
(path/"05").ls()
```

11. Examine one of the images. Note that you may see a different image:

```
img_files = get_image_files(path/'10')
img = PILImage.create(img_files[8])
img.to_thumb(180)
```

Figure 2.19 – One of the images in the BIWI_HEAD_POSE dataset

12. In addition to the image files, this dataset also includes text files that encode the pose depicted in the image. Ingest one of these text files into a pandas DataFrame and display it:

```
df_pose = pd.read_csv(path/'05/frame_00191_pose.txt', header=None)
df_pose.head()
```

	0
0	0.860993 0.162766 -0.481869
1	0.00729371 0.943363 0.331682
2	0.508564 -0.289091 0.811042
3	116.403 12.1671 871.544

Figure 2.20 – The first few records of one of the position text files

In this section, you learned how to ingest two different kinds of image datasets, explore their directory structure, and examine images from the datasets.

How it works...

You used the same `untar_data()` function to ingest the curated tabular, text, and image datasets, and the same `ls()` function to examine the directory structures for all the different kinds of datasets. On top of these common facilities, fastai provides additional convenience functions for examining image data: `get_image_files()` to collect all the image files in a directory tree starting at a given directory, and `to_thumb()` to render the image at a size that is suitable for a notebook.

There's more...

In addition to image classification datasets (where the goal of the trained model is to predict the category of what's displayed in the image) and image localization datasets (where the goal is to predict the location in the image of a given feature), the fastai curated datasets also include image segmentation datasets where the goal is to identify the subsets of an image that contain a particular object, including the CAMVID and CAMVID_TINY datasets.

Cleaning up raw datasets with fastai

Now that we have explored a variety of datasets that are curated by fastai, there is one more topic left to cover in this chapter: how to clean up datasets with fastai. Cleaning up datasets includes dealing with missing values and converting categorical values into numeric identifiers. We need to apply these cleanup steps to datasets because deep learning models can only be trained with numeric data. If we try to train the model with datasets that contain non-numeric data, including missing values and alphanumeric identifiers in categorical columns, the training process will fail. In this section, we are going to review the facilities provided by fastai to make it easy to clean up datasets, and thus make the datasets ready to train deep learning models.

Getting ready

Ensure you have followed the steps in *Chapter 1, Getting Started with fastai*, to get a fastai environment set up. Confirm that you can open the `cleaning_up_datasets.ipynb` notebook in the `ch2` directory of your repository.

How to do it...

In this section, you will be running through the `cleaning_up_datasets.ipynb` notebook to address missing values in the `ADULT_SAMPLE` dataset and replace categorical values with numeric identifiers.

Once you have the notebook open in your fastai environment, complete the following steps:

1. Run the first two cells to import the necessary libraries and set up the notebook for fastai.

2. Recall the *Examining tabular datasets with fastai* section of this chapter. When you checked to see which columns in the `ADULT_SAMPLE` dataset had missing values, you found that some columns did indeed have missing values. We are going to identify the columns in `ADULT_SAMPLE` that have missing values, and use the facilities of fastai to apply transformations to the dataset that deal with the missing values in those columns, and then replace those categorical values with numeric identifiers.

3. First, let's ingest the ADULT_SAMPLE curated dataset again:

```
path = untar_data(URLs.ADULT_SAMPLE)
```

4. Now, create a pandas DataFrame for the dataset and check for the number of missing values in each column. Note which columns have missing values:

```
df = pd.read_csv(path/'adult.csv')
df.isnull().sum()
```

5. To deal with these missing values (and prepare categorical columns), we will use the fastai TabularPandas class (https://docs.fast.ai/tabular. core.html#TabularPandas). To use this class, we need to prepare the following parameters:

    a) **procs** is the list of transformations that will be applied to TabularPandas. Here, we will specify that we want missing values to be filled (FillMissing) and that we will replace values in categorical columns with numeric identifiers (Categorify).

    b) **dep_var** specifies which column is the dependent variable; that is, the target that we want to ultimately predict with the model. In the case of ADULT_SAMPLE, the dependent variable is salary.

    c) **cont** and **cat** are lists of the columns in the dataset. They are continuous and categorical, respectively. Continuous columns contain numeric values, such as integers or floating-point values. Categorical values contain category identifiers, such as names of US states, days of the week, or colors. We use the cont_cat_ split() (https://docs.fast.ai/tabular.core.html#cont_cat_ split) function to automatically identify the continuous and categorical columns:

```
procs = [FillMissing,Categorify]
dep_var = 'salary'
cont,cat = cont_cat_split(df, 1, dep_var=dep_var)
```

6. Now, create a TabularPandas object called df_no_missing using these parameters. This object will contain the dataset with missing values replaced and the values in the categorical columns replaced with numeric identifiers:

```
df_no_missing = TabularPandas(df, procs, cat, cont, y_
names = dep_var)
```

7. Apply the `show` API to `df_no_missing` to display samples of its contents. Note that the values in the categorical columns are maintained when the object is displayed using `show()`. What about replacing the categorical values with numeric identifiers? Don't worry – we'll see that result in the next step:

```
df_no_missing.show(3)
```

	workclass	education	marital-status	occupation	relationship	race	sex	native-country	education-num_na	age
0	Private	Assoc-acdm	Married-civ-spouse	#na#	Wife	White	Female	United-States	False	49
1	Private	Masters	Divorced	Exec-managerial	Not-in-family	White	Male	United-States	False	44
2	Private	HS-grad	Divorced	#na#	Unmarried	Black	Female	United-States	True	38

Figure 2.21 – The first few records of df_no_missing

8. Now, display some sample contents of `df_no_missing` using the `items.head()` API. This time, the categorical columns contain the numeric identifiers rather than the original values. This is an example of a benefit provided by fastai: the switch between the original categorical values and the numeric identifiers is handled elegantly. If you need to see the original values, you can use the `show()` API, which transforms the numeric values in categorical columns back into their original values, while the `items.head()` API shows the actual numeric identifiers in the categorical columns:

```
df_no_missing.items.head(3)
```

	age	workclass	fnlwgt	education	education-num	marital-status	occupation	relationship	race	sex	capital-gain
0	49	5	101320	8	12.0	3	0	6	5	1	0
1	44	5	236746	13	14.0	1	5	2	5	2	10520
2	38	5	96185	12	10.0	1	0	5	3	1	0

Figure 2.22 – The first few records of df_no_missing with numeric identifiers in categorical columns

9. Finally, let's confirm that the missing values were handled correctly. As you can see, the two columns that originally had missing values no longer have missing values in `df_no_missing`:

```
df_no_missing.items.isnull().sum()
age                     0
workclass               0
fnlwgt                  0
education               0
education-num           0
marital-status          0
occupation              0
relationship            0
race                    0
sex                     0
capital-gain            0
capital-loss            0
hours-per-week          0
native-country          0
salary                  0
education-num_na        0
dtype: int64
```

Figure 2.23 – Missing values in df_no_missing

By following these steps, you have seen how fastai makes it easy to prepare a dataset to train a deep learning model. It does this by replacing missing values and converting the values in the categorical columns into numeric identifiers.

How it works...

In this section, you saw several ways that fastai makes it easy to perform common data preparation steps. The `TabularPandas` class provides a lot of value by making it easy to execute common steps to prepare a tabular dataset (including replacing missing values and dealing with categorical columns). The `cont_cat_split()` function automatically identifies continuous and categorical columns in your dataset. In conclusion, fastai makes the cleanup process easy and less error prone than it would be if you had to hand code all the functions required to accomplish these dataset cleanup steps.

3
Training Models with Tabular Data

In the previous chapter, we learned how to ingest various kinds of datasets using fastai and how to clean up datasets. In this chapter, we are going to get into the details of training a model with fastai using tabular data. **Tabular data**, which is data organized in rows and columns that you would find in a spreadsheet file or a database table, is critical to most businesses. The fastai framework acknowledges the importance of tabular data by providing a full suite of features to support deep learning applications based on tabular data.

To explore deep learning with tabular data in fastai, we will return to the `ADULT_SAMPLE` dataset, one of the datasets we examined in *Chapter 2, Exploring and Cleaning Up Data with fastai*. By using this dataset, we will train a deep learning model, while also learning about the `TabularDataLoaders` (used to define the training and test datasets) and `tabular_learner` (used to define and train the model) objects.

We will also look at datasets outside the set of curated datasets to learn how we can ingest non-curated datasets to train deep learning models in fastai. We will wrap up this chapter by exploring what makes a tabular dataset a decent candidate for training a fastai deep learning model and how to save a trained model.

Here are the recipes that will be covered in this chapter:

- Training a model in fastai with a curated tabular dataset
- Training a model in fastai with a non-curated tabular dataset
- Training a model with a standalone dataset
- Assessing whether a tabular dataset is a good candidate for fastai
- Saving a trained tabular model
- Test your knowledge

Technical requirements

Ensure that you have completed the setup sections in *Chapter 1*, *Getting Started with fastai*, and have a working Gradient instance or Colab setup. Ensure that you have cloned the repository for this book (`https://github.com/PacktPublishing/Deep-Learning-with-fastai-Cookbook`) and have access to the `ch3` folder. This folder contains the code samples described in this chapter.

Training a model in fastai with a curated tabular dataset

In *Chapter 2*, *Exploring and Cleaning Up Data with fastai*, you learned how to ingest and examine the `ADULT_SAMPLE` curated tabular dataset. In this recipe, we will go through the process of training a deep learning model on this dataset using fastai. This will give you an overview of the *happy path* to creating a tabular deep learning model with fastai. The goal of this recipe is to use this dataset to train a deep learning model with fastai, which predicts whether the person described in a particular record will have a salary above or below 50k.

Getting ready

Confirm that you can open the `training_with_tabular_datasets.ipynb` notebook in the `ch3` directory of your repository.

I am grateful for the opportunity to include the ADULT_SAMPLE dataset featured in this section.

> **Dataset citation**
>
> Ron Kohavi. (1996) *Scaling Up the Accuracy of Naive-Bayes Classifers: a Decision-Tree Hybrid* (`http://robotics.stanford.edu/~ronnyk/nbtree.pdf`).

How to do it...

In this recipe, you will be running through the `training_with_tabular_datasets.ipynb` notebook. Once you have the notebook open in your fastai environment, complete the following steps:

1. Run the cells in the notebook up to the `Define transforms, dependent variable, continuous and categorical columns` cell. By running these cells, you will be setting up the notebook and ingesting the `ADULT_SAMPLE` curated tabular dataset into a pandas DataFrame, which you will use through the rest of this notebook. These cells are identical to the ones in the notebook shown in *Chapter 2, Exploring and Cleaning Up Data with fastai*, for examining tabular curated datasets `examining_tabular_datasets.ipynb`.

2. Run the first new cell in the notebook with the following code:

    ```
    procs = [FillMissing,Categorify]
    dep_var = 'salary'
    cont,cat = cont_cat_split(df, 1, dep_var=dep_var)
    ```

    This cell sets the following values:

    a) `procs`: This is a list of the transformations that will be applied in the `TabularDataLoaders` object. `FillMissing` specifies that missing the values in a column will be replaced with the median value for the column. `Categorify` specifies that the values in categorical columns will be replaced with numeric identifiers.

    b) `dep_var`: This will be used to identify which column in the dataset contains the dependent variable. This column is also known as the target value or the y value for the model – it is the value that the trained model will predict. For this model, we are predicting the value for the `salary` column.

c) `cont` and `cat`: These are lists of continuous and categorical columns from the `df` DataFrame that are returned by `cont_cat_split`, respectively. This function is a major benefit that fastai provides for tabular deep learning models. It saves a lot of repetitive coding by automatically detecting which columns are continuous (that is, can take on an unlimited set of values, such as currency amounts, physical dimensions, or counts of objects) or categorical (that is, can only take a finite set of distinct values, such as states of the US or days of the week).

3. Run the next cell with the following code to define the `TabularDataLoaders` object called `dls`. Note that some of the parameters specified in the definition of this object (such as batch size) are usually associated with the training process rather than defining the training dataset:

```
dls=TabularDataLoaders.from_df(
df, path, procs= procs,
cat_names= cat, cont_names = cont,
y_names = dep_var,
valid_idx=list(range(1024,1260)), bs=64)
```

The definition of the `TabularDataLoaders` object uses the following arguments:

a) `df, path, procs`: The DataFrame containing the ingested dataset, the path object for the dataset, and the list of transformations defined in the previous step.

b) `cat_names, cont_names`: The lists of categorical and continuous columns defined in the previous step.

c) `y_names`: The column containing the dependent variable/target values.

d) `valid_idx`: The index values of the subset of rows of the `df` DataFrame that will be reserved as the validation dataset for the training process.

e) `bs`: Batch size for the training process.

4. Run the cell with the following code to define and train the model. The first line specifies that the model is being defined using the `TabularDataLoaders` object, which was defined in the previous step with the default number of layers, and using accuracy as the metric that was optimized in the training process. The second line triggers the training process for three epochs; that is, three iterations through the entire dataset:

```
learn = tabular_learner(dls,layers=[200,100],
metrics=accuracy)
learn.fit_one_cycle(3)
```

The output of this cell will be the training results by epoch. The results include the epoch's number, the training loss, the validation loss, and the elapsed time for each epoch:

epoch	train_loss	valid_loss	accuracy	time
0	0.328067	0.376390	0.830508	00:09
1	0.324054	0.356516	0.830508	00:09
2	0.300545	0.340962	0.838983	00:09

Figure 3.1 – Training results

5. Run the following cell to get a sample result of the trained model's predictions. You can compare the `salary` column with the `salary_pred` column (the prediction made by the model – highlighted in the preceding screenshot) to get a snapshot of how the model performed for this sample of rows from the dataset. In this sample set, the model's predictions match the actual values of the dependent variable in the `salary` column:

```
# show sample result, including transformed x, y and predicted transformed y
learn.show_results()
```

	workclass	education	marital-status	occupation	relationship	race	sex	native-country	education-num_na	age	fnlwgt	education-num	capital-gain	capital-loss	hours-per-week	salary	salary_pred
0	7.0	12.0	3.0	13.0	1.0	5.0	2.0	40.0	1.0	58.0	248841.0	9.0	15024.0	0.0	40.0	1.0	1.0
1	8.0	11.0	3.0	11.0	1.0	5.0	2.0	40.0	1.0	41.0	309056.0	16.0	0.0	0.0	50.0	1.0	1.0
2	5.0	5.0	7.0	9.0	3.0	2.0	1.0	31.0	1.0	63.0	106910.0	3.0	0.0	0.0	19.0	0.0	0.0
3	1.0	11.0	3.0	1.0	1.0	5.0	2.0	40.0	1.0	72.0	118902.0	16.0	0.0	2392.0	6.0	1.0	1.0
4	5.0	2.0	5.0	7.0	3.0	3.0	1.0	40.0	1.0	41.0	155657.0	7.0	0.0	0.0	40.0	0.0	0.0
5	5.0	2.0	1.0	8.0	2.0	5.0	2.0	40.0	1.0	38.0	252250.0	7.0	0.0	0.0	65.0	0.0	0.0
6	5.0	16.0	5.0	15.0	5.0	5.0	2.0	40.0	1.0	29.0	277342.0	10.0	0.0	0.0	40.0	0.0	0.0
7	5.0	12.0	3.0	7.0	1.0	5.0	2.0	40.0	1.0	25.0	218667.0	9.0	0.0	0.0	40.0	0.0	0.0
8	7.0	10.0	1.0	5.0	2.0	5.0	2.0	40.0	1.0	54.0	154785.0	13.0	0.0	0.0	50.0	0.0	0.0

Figure 3.2 – Output of show_results

6. Run the following cell to get a summary of the structure of the trained model:

```
learn.summary()
```

The output of this cell includes the following details:

a) A list of all the layers that make up the model:

```
TabularModel (Input shape: ['64 x 9', '64 x 6'])
========================================================
Layer (type)          Output Shape        Param #    Trainable
========================================================
Embedding             64 x 6              60         True

Embedding             64 x 8              136        True

Embedding             64 x 5              40         True

Embedding             64 x 8              128        True
```

Figure 3.3 – List of layers that make up the model

b) The parameters in the trained model, the optimizer and loss functions, and the callbacks used. Callbacks (https://docs.fast.ai/callback.core. html) specify actions to be taken during the training process, such as stopping the training process prior to executing all the epochs. In the case of this trained model, callbacks are automatically specified by fastai to track the number of epochs done (TrainEvalCallback), track losses and metrics by batch/epoch (Recorder), and display the progress bars shown while the model is being trained (ProgressCallBack):

```
Total params: 33,936
Total trainable params: 33,936
Total non-trainable params: 0

Optimizer used: <function Adam at 0x7f1377973820>
Loss function: FlattenedLoss of CrossEntropyLoss()

Model unfrozen

Callbacks:
  - TrainEvalCallback
  - Recorder
  - ProgressCallback
```

Figure 3.4 – Additional details in the output of summary()

You have now completely trained a deep learning model with a curated tabular dataset using fastai.

How it works...

As you saw in this recipe, once the data has been ingested, it only takes a few lines of code with fastai to get a trained deep learning model. The simplicity and compactness of fastai code is partially down to fastai making reasonable default choices when possible.

For example, fastai determines that when the target column is categorical, the model should be predicting choices in a category (in the case of this model, a binary choice of whether a person's salary is above or below 50 k) rather than continuous values. The following are additional benefits provided by fastai for training deep learning models with tabular data:

- Detecting which columns in a DataFrame are categorical or continuous
- Selecting appropriate callbacks for the training process
- Defining the layers for the deep learning model (including embedding layers for the categorical columns)

Let's compare what we would have to do for a Keras deep learning model. In a Keras deep learning model for a tabular dataset, each of these characteristics of the model would have to be explicitly coded, resulting in longer, more complex code to create the model. In addition, somebody learning about deep learning models for tabular datasets would face a bigger challenge by needing to deal with many more details to get their first working trained model. The bottom line is that fastai makes it possible to get to a basic deep learning model faster.

Before moving on to the next recipe, it's worth digging into the model from this recipe because the other recipes in this chapter will follow the same pattern. A deeply detailed description of the model is beyond the scope of this book, so we will just focus on some highlights here.

As shown in the recipe, the model is defined as a `tabular_learner` object (documentation here: `https://docs.fast.ai/tabular.learner.html`). This object is a specialization of the fastai `learner` object which you first saw in the *Understanding the world in four applications: tables, text, recommender systems, and images* section of *Chapter 1, Getting Started with fastai*. You can see the structure of this model from the output of `learn.summary()`. The beginning of the output specifies `Input shape: ['64 x 9', '64 x 6']` - the second dimension of the first element corresponds with the number of categorical columns, and the second dimension of the second element corresponds with the number of continuous columns. There are embedding layers (documentation here: `https://docs.fast.ai/layers.html#Embeddings`) defined for the categorical columns (the columns in the `cat` list) and a `BatchNorm` layer (documentation here: `https://docs.fast.ai/layers.html#BatchNorm-layers`) for the continuous columns.

To see additional details about the structure of the model, see the output of `learn.model`. In particular, you can see that for the final `Linear` layer, `out_features = 2`, corresponds with the binary output of the model (the individual's income is above/below 50 k).

Training a model in fastai with a non-curated tabular dataset

In *Chapter 2, Exploring and Cleaning Up Data with fastai*, you reviewed the curated datasets provided by fastai. In the previous recipe, you created a deep learning model that had been trained on one of these curated datasets. What if you want to train a fastai model for a tabular dataset that is not one of these curated datasets?

In this recipe, we will go through the process of ingesting a non-curated dataset – the Kaggle house prices dataset (`https://www.kaggle.com/c/house-prices-advanced-regression-techniques/data`) – and training a deep learning model on it. This dataset presents some additional challenges. Compared to a curated fastai dataset, there are additional steps required to ingest the dataset, and its structure requires special handling to deal with missing values.

The goal of this recipe is to use this dataset to train a deep learning model, that then predicts whether a house has a sale price that is above or below the average price for the dataset.

Getting ready

To complete this recipe you will need a Kaggle ID. If you don't already have one, you can get one here: `https://www.kaggle.com/account/login`. Once you have your Kaggle ID, complete the following steps to get the token for accessing the Kaggle house prices dataset from within your notebook:

1. Log in with your Kaggle ID, click on your account (top right), and then click on **Account**:

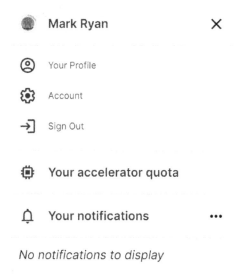

Figure 3.5 – Kaggle account menu

2. On the **Account** page, scroll down to the API section and click on **Create New API Token**. The `kaggle.json` file will be downloaded on your local system:

Figure 3.6 – Selecting Create New API Token to get a new Kaggle API token

3. In your Gradient environment, open a Terminal session, go to the `/root` directory, and create a new directory called `.kaggle`.

4. Upload the `kaggle.json` file that you downloaded in *Step 2* to the new directory that you just created; that is, `/root/.kaggle`.

These steps will prepare you to use the Kaggle house price dataset (`https://www.kaggle.com/c/house-prices-advanced-regression-techniques/data`) in this recipe.

I am grateful for the opportunity to include the house price dataset featured in this section.

> **Dataset citation**
>
> Dean De Cock (2011). Ames, Iowa: Alternative to the Boston Housing Data as an End of Semester Regression Project (`http://jse.amstat.org/v19n3/decock.pdf`) Journal of Statistics Education Volume 19, Number 3(2011), (`www.amstat.org/publications/jse/v19n3/decock.pdf`)

How to do it...

In this recipe, you will be running through the `accessing_non_curated_datasets.ipynb` notebook, as well as the fastai dataset documentation, to understand the datasets that fastai curates. Once you have the notebook open in your fastai environment, complete the following steps:

1. If you have not already done so, install the `kaggle` library by running the following command:

   ```
   pip install kaggle
   ```

2. Run the first three cells of the notebook to load the libraries that you will need for this recipe and prepare the notebook for fastai.

3. In your Gradient environment, go to the `/root/.kaggle` directory and open the `kaggle.json` file. The contents of this file should look like this, with your ID and your 32-character key as the first and second values, respectively:

   ```
   {"username":<YOUR ID>,"key":<YOUR KEY>}
   ```

4. Copy the contents of your `kaggle.json` file.

5. In your copy of the `accessing_non_curated_datasets.ipynb` notebook, paste the content of the `kaggle.json` file into single quotes to assign the value to the variable creds. Then, run the cell:

   ```
   creds = '{"username":<YOUR ID>,"key":<YOUR KEY>}'
   ```

6. Run this cell to set the credential path for the dataset:

```
cred_path = Path('~/.kaggle/kaggle.json').expanduser()
```

7. Run this cell to set the path for your dataset:

```
path = URLs.path('house_price')
```

8. Run this cell to create the target directory for the dataset, download the dataset, unzip the dataset in the target directory, and list the contents of the target directory:

```
if not path.exists():
    path.mkdir()
    api.competition_download_cli('house-prices-advanced-
regression-techniques', path=path)
    file_extract(path/'house-prices-advanced-regression-
techniques.zip')
    path.ls(file_type='text')
```

9. The output of the `path.ls()` function shows the structure of the dataset. In this recipe, we will use `train.csv` to train the deep learning model and then use `test.csv` to exercise the trained model:

```
[Path('/storage/archive/house_price/sample_submission.
csv'),
Path('/storage/archive/house_price/data_description.
txt'),
Path('/storage/archive/house_price/train.csv'),
Path('/storage/archive/house_price/test.csv')]
```

10. Run this cell to ingest the `train.csv` file into a pandas DataFrame called `df_train`:

```
df_train = pd.read_csv(path/'train.csv')
```

11. Run this cell to ingest the `test.csv` file into a pandas DataFrame called `df_test`:

```
df_test = pd.read_csv(path/'test.csv')
```

12. Run the `shape` command to get the dimensions of `df_train` and `df_test`. Notice that `df_test` has one fewer columns than `df_train` – can you think of why this would be the case and which column is missing from `df_test`?

13. Run this cell to define the `under_over()` function, which returns `'0'` if the input value is less than the mean and `'1'` if not:

```
def under_over(x,mean_x):
    if (x <= mean_x):
        returner = '0'
    else:
        returner = '1'
    return(returner)
```

14. Run this cell to use the `under_over()` function, which will replace the values in the `SalePrice` column with indicators of whether the value was above or below the average for the column:

```
mean_sp = int(df_train['SalePrice'].mean())
df_train['SalePrice'] = df_train['SalePrice'].
apply(lambda x: under_over(x,mean_sp))
df_train.head()
```

15. When you display the contents of the `df_train` DataFrame, you will see that the values in the `SalePrice` column have been replaced with zeros and ones:

MoSold	YrSold	SaleType	SaleCondition	SalePrice
2	2008	WD	Normal	1
5	2007	WD	Normal	1
9	2008	WD	Normal	1
2	2006	WD	Abnorml	0
12	2008	WD	Normal	1

Figure 3.7 – The values in the SalePrice column have been replaced

16. Run this cell to see the count of each of the new values in the `SalePrice` column:

```
df_train['SalePrice'].value_counts()
```

17. Run this cell to define the transformation to apply to the dataset, the column that contains the dependent variable (target), and the continuous and categorical columns:

```
dep_var = 'SalePrice'
cont,cat = cont_cat_split(df_train, 1, dep_var=dep_var)
```

This cell sets the following values:

a) `procs`: This is a list of the transformations that will be applied to the `TabularDataLoaders` object. `FillMissing` specifies that the missing values in a column are replaced with the median value for the column. `Categorify` specifies that the values in the categorical columns are replaced with numeric identifiers.

b) `dep_var`: This will be used to identify which column in the dataset contains the dependent variable; that is, the column that contains the value that we want the model to predict. For this model, we are predicting the value for the `SalePrice` column.

c) `cont` and `cat`: These are lists of continuous and categorical columns from the `df_train` DataFrame returned by using `cont_cat_split`.

18. Run this cell to check for missing values in the `df_train` DataFrame. The first line gets a count of missing values in the DataFrame, while the second line defines a new DataFrame, `df_train_missing`, that has a row for each column in the original DataFrame that has at least one missing value. The columns of this DataFrame are the names of the columns with missing values, the missing value count for each column, and the proportion of values that are missing in the column:

```
count = df_train.isna().sum()
df_train_missing = (pd.concat([count.rename('missing_
count'),count.div(len(df_train)).rename('missing_
ratio')],axis = 1).loc[count.ne(0)])
```

Looking at the values in df_train_missing, we can see that some columns have a large number of missing values:

	missing_count	missing_ratio
LotFrontage	259	0.177397
Alley	1369	0.937671
MasVnrType	8	0.005479
MasVnrArea	8	0.005479
BsmtQual	37	0.025342

Figure 3.8 – Rows from df_train_missing

19. Run this cell to deal with the missing values in the df_train and df_test DataFrames. The first two statements replace the missing values in the categorical columns with the most common non-missing value, while the second two statements replace the missing values in the continuous columns with zeros:

```
df_train[cat] = df_train[cat].fillna(df_train[cat].
mode().iloc[0])
df_test[cat] = df_test[cat].fillna(df_test[cat].mode().
iloc[0])

df_train[cont] = df_train[cont].fillna(0.0)
df_test[cont] = df_test[cont].fillna(0.0)
```

20. Run this cell to check for missing values in the df_train DataFrame once more:

```
count = df_train.isna().sum()
df_train_missing = (pd.concat([count.rename('missing_
count'),count.div(len(df_train)).rename('missing_
ratio')],axis = 1).loc[count.ne(0)])
```

21. Now, when you check the contents of the df_train_missing DataFrame, it will be empty, confirming that all the missing values have been dealt with:

df_train_missing

missing_count   missing_ratio

Figure 3.9 – Confirmation that the missing values have been dealt with

22. Run this cell to create a `TabularDataLoaders` object. The first line defines the transformation procedures to be applied in the `TabularDataLoaders` object, while the second line defines the `TabularDataLoaders` object:

```
procs = [Categorify, Normalize]
dls_house=TabularDataLoaders.from_df(
    df_train,path,procs= procs,
    cat_names= cat, cont_names = cont, y_names = dep_var,
    valid_idx=list(range((df_train.shape[0]-100),df_
train.shape[0])),
    bs=64)
```

Here are the arguments for the definition of the `TabularDataLoaders` object:

a) `procs`: This is a list of the transformations that will be applied to the `TabularDataLoaders` object. `Normalize` specifies that the values are all scaled to a consistent range. `Categorify` specifies that the values in the categorical columns are replaced with numeric identifiers.

b) `df_train, path, procs`: The DataFrame containing the ingested dataset, the path object for the dataset, and the list of transformations defined in the previous step.

c) `cat_names, cont_names`: The lists of categorical and continuous columns.

d) `y_names`: The column containing the dependent variable/target values.

e) `valid_idx`: The index values of the subset of rows of the `df` DataFrame that will be reserved as the validation dataset for the training process.

f) `bs`: The batch size for the training process.

23. Run this cell to define and train the deep learning model. The first line specifies that the model is being created using the `TabularDataLoaders` object we defined in the previous step, with the default number of layers and accuracy as the metrics being optimized in the training process. The second line triggers the training process for 5 epochs:

```
learn = tabular_learner(dls_house, layers=[200,100],
metrics=accuracy)
learn.fit_one_cycle(5)
```

The training process produces an output that shows the training loss, validation loss, and accuracy for each epoch. This means we have trained a model that can predict whether the cost of a given property from the validation set is above or below the average with 92% accuracy. Can you think of some reasons why the accuracy of this model is somewhat better than the accuracy of the model you trained in the previous recipe?

epoch	train_loss	valid_loss	accuracy	time
0	0.524770	0.579092	0.820000	00:00
1	0.348470	0.325142	0.870000	00:00
2	0.248136	0.216792	0.920000	00:00
3	0.194248	0.159855	0.920000	00:00
4	0.156570	0.166303	0.920000	00:00

Figure 3.10 – Results of training on the non-curated dataset

24. You can run this cell to apply the trained model to the test dataset. Note that because of the structure of this dataset, the test dataset does not include y-dependent values, which means that while you can apply the model to the test records, you don't have any way of assessing the accuracy of the predictions that the model makes on the test set:

```
dl = learn.dls.test_dl(df_test)
```

25. Run this cell to get a sample of predictions that the trained model makes on the test dataset:

```
learn.show_results()
```

The output of `learn.show_results()` lets you see the result of applying the trained model to the dataset:

Porch	3SsnPorch	ScreenPorch	PoolArea	MiscVal	MoSold	YrSold	SalePrice	SalePrice_pred
52508	-0.117549	-0.271653	-0.06042	-0.085622	-0.117503	0.146225	1.0	1.0
52508	-0.117549	-0.271653	-0.06042	-0.085622	0.622329	-0.607057	0.0	0.0
31317	-0.117549	-0.271653	-0.06042	-0.085622	-0.487419	-0.607057	0.0	0.0
52508	-0.117549	-0.271653	-0.06042	-0.085622	1.362161	-1.360339	1.0	1.0
52508	-0.117549	-0.271653	-0.06042	-0.085622	0.622329	-1.360339	0.0	0.0
75433	-0.117549	-0.271653	-0.06042	-0.085622	-0.117503	1.652790	0.0	0.0

Figure 3.11 – The subset of results from applying the model trained to a non-curated dataset

You have now gone through the process of using a non-curated dataset to train a fastai deep learning model.

How it works...

In this recipe, you learned how to adapt to a different kind of dataset where the train and test data is separate. In this case, you cannot rely on the transformations in the `TabularDataLoaders` object to deal with missing data. That's why the code associated with this recipe deals with the missing values in each of the train and test datasets individually.

Training a model with a standalone dataset

In the previous recipes in this chapter, we looked at training fastai models on a curated tabular dataset and a dataset directly loaded from Kaggle. In this recipe, we are going to examine how to train a model with a dataset that is from a self-standing file. The dataset we will use in this recipe is made up of property listings in Kuala Lumpur, Malaysia and is available from the Kaggle site at `https://www.kaggle.com/dragonduck/property-listings-in-kuala-lumpur`.

This dataset is not like the tabular datasets we have seen so far. The datasets we have already encountered have been well-behaved and have only required a small amount of cleanup. The Kualu Lumpur property dataset, by contrast, is a real-world dataset. In addition to missing values, it contains many errors and irregularities. It is also large enough (over 50k records) to give deep learning a decent chance to be useful on it.

Getting ready

Ensure you have followed the steps in *Chapter 1, Getting Started with fastai*, so that you have a fastai environment set up. Confirm that you can open the `training_model_standalone_tabular_dataset.ipynb` notebook in the `ch3` directory of your repository. Also, ensure that you have uploaded the data file by following these steps:

1. Download `data_kaggle.csv.zip` from `https://www.kaggle.com/dragonduck/property-listings-in-kuala-lumpur`.

2. Unzip the downloaded file to extract `data_kaggle.csv`.

3. From the Terminal in your Gradient environment, make your current directory `/storage/archive`:

    ```
    cd /storage/archive
    ```

4. Create a folder called `/storage/archive/kl_property`:

    ```
    mkdir kl_property
    ```

5. Upload `data_kaggle.csv` to `/storage/archive/kl_property`. You can use the upload button in JupyterLab in Gradient to do the upload, but you need to do so by performing several steps:

    a) From the Terminal in your Gradient environment, make `/notebooks` your current directory:

    ```
    cd /notebooks
    ```

    b) Make a new directory called `/notebooks/temp`:

    ```
    mkdir temp
    ```

c) In the JupyterLab file browser, make `temp` your current folder, select the upload button, as shown in the following screenshot, and select the `data_kaggle.csv` file from your local system folder where you extracted it in *step 2*:

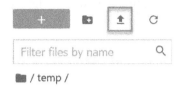

Figure 3.12 – Upload button in JupyterLab

d) From the Terminal in your Gradient environment, copy `data_kaggle.csv` into `/storage/archive/kl_property`:

```
cp /notebooks/temp/data_kaggle.csv /storage/archive/kl_
property/data_kaggle.csv
```

I want to acknowledge the dataset featured in this section and express my gratitude for the opportunity to include it in the book.

> **Dataset citation**
>
> Jas S (2019). *Property Listings in Kuala Lumpur* (`https://www.kaggle.com/dragonduck/property-listings-in-kuala-lumpur`)

Let's see how to go about it in the next section.

How to do it...

In this recipe, you will be running through the `training_model_standalone_tabular_dataset.ipynb` notebook to train a model using the Kuala Lumpur property prices dataset.

Once you have the notebook open in your fastai environment, follow these steps:

1. Run the first three cells to import the necessary libraries and set up the notebook for fastai.

2. Run this cell to associate `path` with the directory that you copied the `data_kaggle.csv` file into:

```
path = URLs.path('kl_property')
```

3. Run this cell to ingest the dataset into the df_train DataFrame:

    ```
    df_train = pd.read_csv(path/'data_kaggle.csv')
    ```

4. Run the following cell to see the first few rows of the dataset:

    ```
    df_train.head()
    ```

    The output of df_train.head() shows a sample of rows from the dataset:

	Location	Price	Rooms	Bathrooms	Car Parks	Property Type	Size	Furnishing
0	KLCC, Kuala Lumpur	RM 1,250,000	2+1	3.0	2.0	Serviced Residence	Built-up : 1,335 sq. ft.	Fully Furnished
1	Damansara Heights, Kuala Lumpur	RM 6,800,000	6	7.0	NaN	Bungalow	Land area : 6900 sq. ft.	Partly Furnished
2	Dutamas, Kuala Lumpur	RM 1,030,000	3	4.0	2.0	Condominium (Corner)	Built-up : 1,875 sq. ft.	Partly Furnished
3	Cheras, Kuala Lumpur	NaN	NaN	NaN	NaN	NaN	NaN	NaN
4	Bukit Jalil, Kuala Lumpur	RM 900,000	4+1	3.0	2.0	Condominium (Corner)	Built-up : 1,513 sq. ft.	Partly Furnished

Figure 3.13 – A sample of rows from the Kuala Lumpur property dataset

5. Note the values in the Price column. The goal of this recipe is to train a deep learning model to predict whether values in this column are above or below the average. To do this, we need to start with numeric values in this column. You can see three problems just from this small sample of data:

    a) The values include RM, the symbol for the ringgit, the Malaysian currency.

    b) The values include a comma thousand separator.

    c) Some rows have a missing value in this column.

    Over the next few cells of the notebook, we will deal with these problems so that we end up with this column containing entirely valid numeric values.

6. Run this cell to get the number of rows in the dataset. The first element of the output is the number of rows in the DataFrame:

    ```
    df_train.shape
    ```

7. Run the cells to define the remove_currency() and remove_after_space() functions. You will need these functions to clean up the dataset.

8. Run this cell to address the problems with the `Price` column demonstrated by the output of `df_train.head()`. This cell drops rows in the dataset where the `Price` value is missing (first statement), removes the currency symbol from the `Price` column (second statement), and converts the values in the `Price` column into numeric values after removing any commas:

```
df_train.dropna(subset=['Price'], inplace=True)
df_train['Price'] = df_train['Price'].apply(lambda x:
remove_currency("RM ",x))
df_train['Price'] = pd.to_numeric(df_train['Price'].str.
replace(',',''), errors='coerce')
```

9. Run this cell to get the number of rows in the dataset again. It's important to know this number because we have eliminated some rows from the dataset in the last cell, and we will remove more as we clean up the `Size` column. By checking the shape of the DataFrame as we are removing rows, we can be sure we're not losing too much information:

```
df_train.shape
```

```
(53635, 8)
```

Figure 3.14 – Getting the shape of the DataFrame prior to dropping rows

10. The `Size` column has a lot of useful information in it, but we have to do some work to get it ready to help train a deep learning model. To start with, run the following cell to see some examples of the values in the `Size` column. Here is the information we want to extract from this column:

a) Extract the prefix (for example, `Built-up`, `Land area`, and so on) into a new column.

b) Get a single numeric value for the remainder of the contents of the column; that is, replace 1,335 sq. ft. with 1,335 and replace 22 x 80 sq. ft. with 1,760.

c) Drop rows where the suffix in the `Size` column cannot yield a numeric value.

The output of df_train['Size'].head() shows examples of the values in the Size column:

```
df_train['Size'].head(10)

0            Built-up : 1,335 sq. ft.
1            Land area : 6900 sq. ft.
2            Built-up : 1,875 sq. ft.
4            Built-up : 1,513 sq. ft.
5            Land area : 7200 sq. ft.
7            Land area : 3600 sq. ft.
8          Land area : 25x75 sq. ft.
9              Built-up : 904 sq. ft.
11     Land area : 22 x 80 sq. ft.
12         Land area : 1900 sq. ft.
Name: Size, dtype: object
```

Figure 3.15 – Examples of values in the Size column

Here are the operations we want to perform on the Size column to prepare it to train a deep learning model:

a) Extract the prefix (for example, Built-up, Land area, and so on) into a new column.

b) Where possible, get a single numeric value for the remainder of the contents of the Size column. For example, we want to replace 1,335 sq. ft. with 1,335 and replace 22 x 80 sq. ft. with 1,760.

c) For rows where it is not possible to get a numeric value from the remainder of the contents of the Size column, drop the row.

11. Run the cell to define the clean_up_size() function. You will use this function to perform the following set of cleanup steps on the Size column. The result will be a DataFrame where all values in the Size column are numeric values representing the area of the property. Here are some of the transformations that are performed by the clean_up_size() function:

a) Lowercase all the values in the Size column.

b) Split the Size column into a new column (Size_type) that contains the non-numeric information and a remainder Size column that contains the numeric information about the area of the property.

c) Replace the missing values in the new `Size` column with `0`.

d) Remove the rows that do not contain any digits.

e) Remove the rows that contain problematic substrings, as listed in `clean_up_list`.

f) Replace extraneous characters so that all the `Size` entries are either numeric or of the `numerica * numericb` form.

g) Replace values of the `numerica * numericb` form with the product of the two values; that is, `numerica` and `numericb`.

12. Run this cell to execute the `clean_up_size()` function:

```
clean_up_list = ["-","\+",'\'','\~',"xx","sf","acre","#"]
df_train = clean_up_size(df_train,clean_up_list)
```

13. Run the `shape` command again to get the shape of the DataFrame after dropping the rows that don't contain enough data to be useful. This confirms that we lost about 2% of the rows from the original dataset after removing the offending rows:

```
# get the record count after the Size column has been cleaned up
df_train.shape
```

```
(52309, 9)
```

Figure 3.16 – Getting the shape of the DataFrame after cleanup

14. Run `df_train.head()` to see a sample of the DataFrame after the cleanup steps. Note that now, we have the following:

a) There is a new `Size_type` column.

b) The `Size` column contains numeric values.

The output of df_train.head() shows what the DataFrame looks like after the Price and Size columns have been cleaned up:

df_train.head()

	Location	Price	Rooms	Bathrooms	Car Parks	Property Type	Size	Furnishing	Size_type
0	KLCC, Kuala Lumpur	1250000	2+1	3.0	2.0	Serviced Residence	1335.0	Fully Furnished	built-up
1	Damansara Heights, Kuala Lumpur	6800000	6	7.0	NaN	Bungalow	6900.0	Partly Furnished	land area
2	Dutamas, Kuala Lumpur	1030000	3	4.0	2.0	Condominium (Corner)	1875.0	Partly Furnished	built-up
4	Bukit Jalil, Kuala Lumpur	900000	4+1	3.0	2.0	Condominium (Corner)	1513.0	Partly Furnished	built-up
5	Taman Tun Dr Ismail, Kuala Lumpur	5350000	4+2	5.0	4.0	Bungalow	7200.0	Partly Furnished	land area

Figure 3.17 – The DataFrame after performing the cleanup steps on the Price and Size columns

15. Run the cell to define the under_over() function. You will run this function to replace the values in the Price column with an indicator of whether the price is under or over the average price.

16. Run the following cell to replace the values in the Price column with indicators of whether the price is above or below the average:

```
mean_sp = int(df_train['Price'].mean())
if categorical_target:
    df_train['Price'] = df_train['Price'].apply(lambda x:
under_over(x,mean_sp))
```

After running this cell, the values in the Price column will be replaced:

	Location	Price	Rooms	Bathrooms	Car Parks	Property Type	Size	Furnishing	Size_type
0	KLCC, Kuala Lumpur	0	2+1	3.0	2.0	Serviced Residence	1335.0	Fully Furnished	built-up
1	Damansara Heights, Kuala Lumpur	1	6	7.0	NaN	Bungalow	6900.0	Partly Furnished	land area
2	Dutamas, Kuala Lumpur	0	3	4.0	2.0	Condominium (Corner)	1875.0	Partly Furnished	built-up
4	Bukit Jalil, Kuala Lumpur	0	4+1	3.0	2.0	Condominium (Corner)	1513.0	Partly Furnished	built-up
5	Taman Tun Dr Ismail, Kuala Lumpur	1	4+2	5.0	4.0	Bungalow	7200.0	Partly Furnished	land area

Figure 3.18 – The DataFrame after replacing the Price values with under/over average indicators

17. Run this cell to define the transformations to be applied to
    `TabularDataLoaders` object, the target column (`Price`), and the continuous
    and categorical column lists:

```
procs = [FillMissing,Categorify]
dep_var = 'Price'
cont,cat = cont_cat_split(df_train, 1, dep_var=dep_var)
```

18. Run this cell to define the `TabularDataLoaders` object with the arguments that
    you have defined so far in the notebook, including the dataset (`df_train`), the
    list of transformations (`procs`) to be applied to the dataset, the continuous and
    categorical column lists (`cont` and `cat`), and the dependent variable (`dep_var`):

```
dls = TabularDataLoaders.from_df(df_train,path,procs=
procs,
           cat_names= cat, cont_names = cont,
           y_names = dep_var,
           valid_idx=list(range((df_train.shape[0]-
5000),df_train.shape[0])),
           bs=64)
```

19. Run this cell to fit the model and see the model's performance. Your accuracy and
    loss may be slightly different, but you should see over 90% accuracy, which is good
    for a model that's been trained on fewer than 100k records:

```
# define and fit the model
learn = tabular_learner(dls, metrics=accuracy)
learn.fit_one_cycle(3)
```

epoch	train_loss	valid_loss	accuracy	time
0	0.175031	0.174658	0.929600	00:10
1	0.143417	0.147668	0.936000	00:09
2	0.128981	0.146698	0.938600	00:10

Figure 3.19 – Results of fitting the model

20. Run this cell to see the results in the validation set. You will see that for this set of results, the model correctly predicts whether a property will have a price above or below the average:

```
# show a set of results from the model
learn.show_results()
```

	Location	Rooms	Property Type	Furnishing	Size_type	Bathrooms_na	Car Parks_na	Bathrooms	Car Parks	Size	Price	Price_pred
0	3.0	0.0	56.0	4.0	2.0	2.0	2.0	-0.053581	-0.010860	0.245799	1.0	1.0
1	14.0	1.0	88.0	2.0	1.0	1.0	1.0	-1.299513	-0.931626	-0.042105	0.0	0.0
2	56.0	2.0	88.0	2.0	1.0	1.0	1.0	-1.299513	-0.931626	-0.027924	0.0	0.0
3	89.0	27.0	76.0	0.0	2.0	2.0	2.0	-0.053581	-0.010860	0.947749	1.0	1.0
4	25.0	0.0	74.0	0.0	2.0	2.0	2.0	-0.053581	-0.010860	0.150699	1.0	1.0
5	19.0	24.0	51.0	2.0	2.0	1.0	2.0	0.569386	-0.010860	0.525406	1.0	1.0
6	14.0	18.0	85.0	1.0	1.0	1.0	2.0	-0.053581	-0.010860	-0.026419	1.0	1.0
7	45.0	18.0	84.0	1.0	1.0	1.0	1.0	-0.676547	-0.931626	-0.035274	0.0	0.0
8	56.0	1.0	84.0	0.0	1.0	1.0	1.0	-0.676547	-0.931626	-0.039847	0.0	0.0

Figure 3.20 – Model predictions of whether properties will have a price above or below the average

Congratulations! You have trained a deep learning model with fastai on a dataset that required non-trivial cleanup before it could be used to train the model.

How it works...

In this recipe, you saw that while fastai provides utilities that make it easier to train a deep learning model on a tabular dataset, you still have to ensure that the dataset is capable of training the model. This means that non-numeric values need to be removed from numeric columns. This may require iterative cleanup steps, as you saw while working through the notebook featured in this recipe.

The dataset featured in this recipe contains the kind of anomalies and inconsistencies that are typical of real-world datasets, so the techniques you exercised in this recipe (including pandas manipulations to remove rows that had problematic values in certain columns, as well as string replacement techniques) will be applicable to other real-world datasets.

Even with a messy dataset, fastai makes it easy to get a high performing deep learning model, thanks to it picking intelligent defaults and automating key operations (such as identifying categorical and continuous columns and dealing with missing values).

Assessing whether a tabular dataset is a good candidate for fastai

So far in this chapter, we have created three deep learning models for tabular datasets using fastai. But what if you want to determine whether a new dataset is a good candidate for training a deep learning model with fastai? In this recipe, we'll go through the process of assessing whether a dataset is a good candidate for deep learning with fastai.

Getting ready

Ensure you have followed the steps in *Chapter 1, Getting Started with fastai,* to get a fastai environment set up.

How to do it...

As you have seen so far in this chapter, you have many choices surrounding datasets that could possibly be applied to deep learning. To assess whether a dataset is a good candidate, we will go through the process of creating a new notebook from scratch and ingesting data from an online API. Follow these steps:

1. Create a new notebook in Gradient. You can do this in Gradient JupyterLab by following these steps:

    a) Click on the new launcher button (+) in the main window:

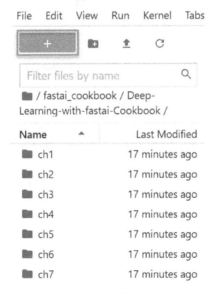

Figure 3.21 – Getting a new launcher in JupyterLab

b) When the launcher pane opens, click on **Python 3** to open a new notebook:

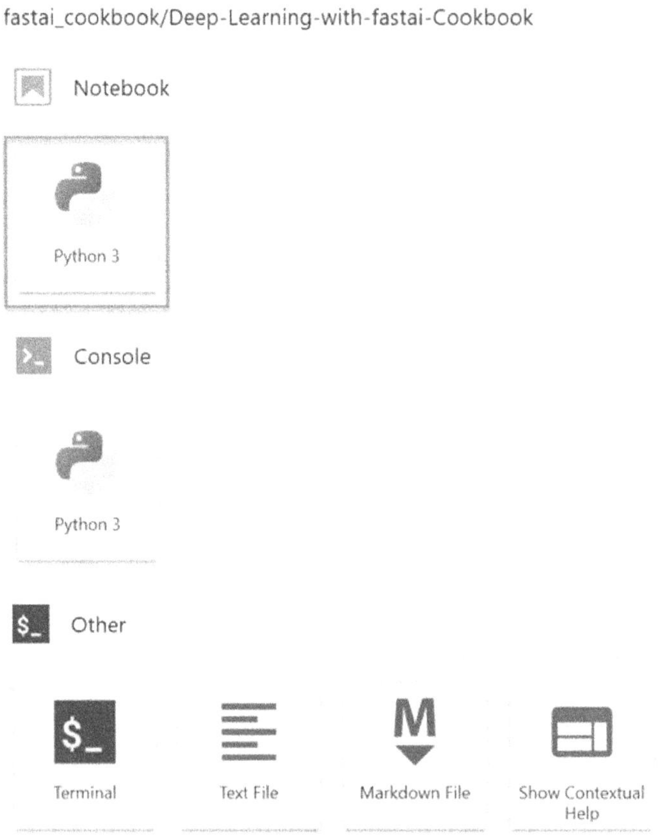

Figure 3.22 – The launcher pane in JupyterLab

2. In the new notebook, create and run two new cells with the statements required to set up the notebook:

```
# imports for notebook boilerplate
!pip install -Uqq fastbook
import fastbook
from fastbook import *
from fastai.tabular.all import *


# set up the notebook for fast.ai
fastbook.setup_book()
```

Figure 3.23 – The cells to set up a fastai notebook for tabular datasets

3. Run the following cell to import the additional libraries required to investigate this dataset:

```
! pip install pandas_datareader
import numpy as np
import pandas as pd
import os
import yaml
# For reading stock data
from pandas_datareader.data import DataReader


# For time stamps
from datetime import datetime
```

4. Run the following cell to load a dataset of stock prices for the company AstraZeneca (stock ticker = AZN):

```
df = DataReader('AZN', 'stooq')
```

5. Check the output of df.head() to see the contents of the dataframe:

	Open	High	Low	Close	Volume
Date					
2021-07-09	59.12	59.910	59.070	59.63	6173577
2021-07-08	59.45	59.670	58.830	59.26	6894951
2021-07-07	59.76	59.820	59.295	59.76	6434879
2021-07-06	60.41	60.555	59.870	59.90	6893637
2021-07-02	60.26	60.875	60.065	60.79	5198428

Figure 3.24 – Sample of the stock prices dataset

6. Get the shape of the dataframe df to determine how many rows are in the dataframe. Do you think this dataset is big enough to successfully train a deep learning model?

```
df.shape

(1258, 5)
```

Figure 3.25 – Getting the shape of df

7. Run the following cell to prepare to check the dataframe for missing values

```
count = df.isna().sum()
df_missing = (pd.concat([count.rename('missing_
count'),count.div(len(df)).rename('missing_ratio')],axis
= 1).loc[count.ne(0)])
```

8. Now confirm the dataset has no missing values

```
# check for missing values
df_missing
```

missing_count   missing_ratio

Figure 3.26 – Confirming the dataset has no missing values

9. Run the following cell to set the parameters for the training run:

```
dep_var = 'Close'
# define columns that are continuous / categorical
cont,cat = cont_cat_split(df, 1, dep_var=dep_var)
```

10. Run the following cell to define the TabularDataLoaders object::

```
procs = [Normalize]
dls = TabularDataLoaders.from_df(df,procs= procs,
        cat_names= cat, cont_names = cont,
        y_names = dep_var,
        valid_idx=list(range((df.shape[0]-50),df.
shape[0])), bs=64)
```

11. Run the following cell to define and train the model::

```
learn = tabular_learner(dls, metrics=accuracy)
learn.fit_one_cycle(30)
```

12. You can see from the following output that the performance of the model is poor:

epoch	train_loss	valid_loss	accuracy	time
0	1562.704834	698.903931	0.000000	00:00
1	1393.395386	487.360260	0.000000	00:00
2	1241.338989	416.841797	0.000000	00:00

Figure 3.27 – Poor performance when training the model

13. Now we want to make some changes to try to get a model that has better performance. To start with, run the following cell to define a function to create a new target column:

```
def get_target(value,threshold):
    '''return based on whether the input value is greater
than or less than input threshold'''
    if value <= threshold:
        return_value = "0"
    else:
        return_value = "1"
    return(return_value)
```

14. Run the following cell to define the new target column:

```
threshold = df['Close'].mean()
df['target'] = df['Close'].apply(lambda x: get_
target(x,threshold))
```

15. Run the following cell to specify that the new `target` column is the dependent variable and to limit `cont`, the set of continuous columns used to train the model. Note that in the first model in this section, the dependent variable was `Close`, a continuous column. That means the first model was trying to predict a continuous value. Because `target` is a categorical column, the new model will predict a categorical value rather than a continuous value:

```
dep_var = 'target'
cont = ['High', 'Low', 'Open', 'Volume']
```

16. Run the following cell to train a new model using the new dependent variable and the new set of continuous columns:

```
dls = TabularDataLoaders.from_df(df,procs= procs,
    cat_names= cat, cont_names = cont,
    y_names = dep_var,
    valid_idx=list(range((df.shape[0]-50),df.shape[0])),
    bs=64)
learn = tabular_learner(dls, metrics=accuracy)
learn.fit_one_cycle(30)
```

17. You can see that this new model has much better performance than the previous model:

epoch	train_loss	valid_loss	accuracy	time
0	0.645138	0.602366	0.980000	00:00
1	0.520773	0.289147	1.000000	00:00
2	0.383414	0.065214	1.000000	00:00

Figure 3.28 – Improved performance with the second model

In this recipe, you tried two variations on a deep learning model for a dataset of stock price information. The first model had a continuous dependent variable and used fastai defaults throughout. Unlike the other recipes in this chapter, the first model had poor performance. If you attempt to train the first model with more epochs you will see that the performance does not improve. The second model, where the dependent variable is changed from continuous to categorical, has much better performance.

How it works...

The first model in this section was unsuccessful. Attempting to predict a continuous value with a deep learning model trained on 1.3 k records is not likely to work. Generally speaking, you need a training set that is an order of magnitude bigger, in the hundreds of thousands or millions of records, to predict a continuous outcome.

Saving a trained tabular model

So far, we have trained a series of fastai deep learning models on tabular datasets. These models are available to us in the Python session where we train the model, but what can we do to save the models so that we can use them later in a different session? In this recipe, we will learn how to save a fastai deep learning model to a file and access that model in another Python session.

Getting ready

Ensure you have followed the steps in *Chapter 1, Getting Started with fastai*, to get a fastai environment set up. Confirm that you can open the `saving_models_trained_with_tabular_datasets.ipynb` and `loading_saved_models_trained_with_tabular_datasets.ipynb` notebooks in the `ch3` directory of your repository.

How to do it...

In this recipe, you will be running through the `saving_models_trained_with_tabular_datasets.ipynb` notebook to train a model – the same model that you trained in the first recipe of this chapter – and save it. Then, you will use the `loading_saved_models_trained_with_tabular_datasets.ipynb` notebook to load and exercise a saved fastai model.

Once you have the `saving_models_trained_with_tabular_datasets.ipynb` notebook open in your fastai environment, follow these steps:

1. Run the cells in the notebook up to the `Save the trained model` cell. By running these cells, you will be ingesting the `ADULT_SAMPLE` curated tabular dataset into a pandas DataFrame and training a fastai model on it.

2. Run the next two cells to set the value of the path for the model to a writable directory. Ensure that the directory that you set learn.path to exists and is writable.

3. Run this cell to save the trained model to the `adult_sample_model.pkl` file:

    ```
    learn.export('adult_sample_model.pkl')
    ```

4. Now that you have saved the trained model into a file, you must load it into another notebook to test the process you would go through to retrieve the model in a new Python session, and then use the saved model to make a prediction on the test data.

5. Open the `loading_saved_models_trained_with_tabular_datasets.ipynb` notebook in a Gradient session.

6. Run the cells up to the `Load the saved, trained model` cell to load the required libraries and set up the notebook.

7. Run this cell to load the model you saved earlier in this recipe into this new notebook. Ensure that you specify the path where you saved the model earlier in this recipe:

    ```
    learn = load_learner('/notebooks/temp/models/adult_
    sample_model.pkl')
    ```

8. Run this cell to load the test dataset into a DataFrame:

    ```
    df_test = pd.read_csv('adult_sample_test.csv')
    ```

9. Run this cell to select the first row of the test dataset and apply the trained model to get a prediction for this data point:

    ```
    test_sample = df_test.iloc[0]
    learn.predict(test_sample)
    ```

10. The result includes the prediction of the model on this data point. You can see that for this data point, the model is predicting a `salary` value of `1.0`, which means that it is predicting that this individual will have a salary of over 50k:

```
(   workclass   education   marital-status   occupation   relationship   race   sex   \
0         5.0         8.0              3.0          0.0            6.0    5.0   1.0

    native-country   education-num_na    age      fnlwgt   education-num   \
0             40.0                1.0   49.0   101320.0            12.0

    capital-gain   capital-loss   hours-per-week   salary
0            0.0         1902.0             40.0      1.0   ,
tensor(1),
tensor([0.2312, 0.7688]))
```

Figure 3.29 – Results of applying the saved model to a test data point

Congratulations! You have successfully saved a fastai deep learning model, loaded the saved model in a new notebook, and applied the saved model to get a prediction on a row of test data.

How it works...

The fastai framework includes support to make it easy to save deep learning models to your filesystem using the `export()` method of `learner` objects. In this recipe, you saw an example of how you can save a trained model to a pickle file. You also learned how to load the pickle file back into Python and then apply the trained model to a new data example. This is a peek ahead at the process of performing inference on a deployed model. In *Chapter 7, Deployment and Model Maintenance*, you will see complete examples of performing inference on a deployed model.

Test your knowledge

Now that you have completed the recipes in this chapter, follow the steps shown here to exercise what you have learned. You will do this by adapting one of the notebooks you worked through in this chapter so that it works with a new dataset.

Getting ready

Follow these steps to upload a new tabular dataset:

1. Go to the site for the Kaggle competition on future sales prediction (`https://www.kaggle.com/c/competitive-data-science-predict-future-sales/data`) and accept the conditions for the competition to get access to the datasets associated with the competition.

2. Download the `sales_train.csv.zip` and `test.csv.zip` files.

3. Unzip the downloaded files to extract `sales_train.csv` and `test.csv`.

4. From the Terminal in your Gradient environment, make your current directory `/storage/archive`:

   cd /storage/archive

5. Create a folder called `/storage/archive/price_prediction`:

   mkdir price_prediction

6. Upload `sales_train.csv` and `test.csv` to `/storage/archive/price_prediction`. You can use the upload button in JupyterLab in Gradient to do the upload via the `/notebooks/temp` directory that you created earlier in this chapter:

   a) In the JupyterLab file browser, make `temp` your current folder, select the upload button, as shown in the following screenshot, and select `sales_train.csv` and `test.csv` from the local system folder where you extracted them in *step 2*:

Figure 3.30 – Upload button in JupyterLab

   b) From the Terminal in your Gradient environment, make `/storage/archive/price_prediction` your current directory:

   cd /storage/archive/price_prediction

c) Copy `sales_train.csv`, `test.csv`, and `data_kaggle.csv` into `/storage/archive/price_prediction`:

```
cp /notebooks/temp/sales_train.csv  sales_train.csv
cp /notebooks/temp/test.csv  test.csv
```

Now that you have uploaded the dataset, it's time to create a notebook to ingest the dataset:

1. Make a copy of the `training_model_standalone_tabular_dataset.ipynb` notebook that you worked through in the *Training a model with a standalone dataset* recipe. Call the copy `training_model_new_tabular_dataset.ipynb`.

2. In your new notebook, update the cells that ingest the training dataset:

```
path = URLs.path('price_prediction')
df_train = pd.read_csv(path/'sales_train.csv')
```

3. The output of `df_train.head()` should show you the structure of the dataset. You can find a description of the columns of this dataset at `https://www.kaggle.com/c/competitive-data-science-predict-future-sales/data?select=sales_train.csv`:

```
df_train.head()
```

	date	date_block_num	shop_id	item_id	item_price	item_cnt_day
0	02.01.2013	0	59	22154	999.00	1.0
1	03.01.2013	0	25	2552	899.00	1.0
2	05.01.2013	0	25	2552	899.00	-1.0
3	06.01.2013	0	25	2554	1709.05	1.0
4	15.01.2013	0	25	2555	1099.00	1.0

Figure 3.31 – A sample of the contents of sales_train.csv

Congratulations! You have ingested another standalone tabular dataset into a fastai notebook. You can apply similar techniques to make other tabular datasets available so that they can used in fastai solutions.

Now that you have ingested the dataset, consider the steps you would take to prepare this dataset for training a deep learning model. How would you deal with any missing values? Are there tests you can apply to some of the columns to detect and correct incorrect values? The Kaggle competition predicts the total sales for every product and store for the next month. Completing a fastai model to tackle this problem is beyond the scope of this book, but consider how you might refactor the dataset to prepare it for this problem.

4
Training Models with Text Data

In *Chapter 3*, *Training Models with Tabular Data*, you went through a series of recipes that demonstrated how to use the facilities of fastai to train deep learning models on tabular data. In this chapter, we will examine how to take advantage of the fastai framework to train deep learning models on text datasets.

To explore deep learning with text data in fastai, we will start by taking a pre-trained **language model** (that is, a model that, when given a phrase, predicts what words come next) and fine-tuning it with the IMDb curated dataset. We will then use the resulting fine-tuned language model to create a **text classifier model** for the movie review use case represented by the IMDb dataset. The text classifier predicts the class of a phrase; in the movie review use case, it predicts whether a given phrase is **positive** or **negative**.

Finally, we apply the same approach to a standalone (that is, non-curated) text dataset of Covid-related tweets. First, we will fine-tune the existing language model on the Covid tweets dataset. Then, we will use the fine-tuned language model to train a text classifier that predicts the class of a phrase according to the categories defined in the Covid tweets dataset: **extremely negative**, **negative**, **neutral**, **positive**, and **extremely positive**.

The approach to training deep learning models on text datasets that is used in this chapter, also known as **ULMFiT**, was initially described by the creators of fastai in their paper entitled *Universal Language Model Fine-Tuning for Text Classification* `https://arxiv.org/abs/1801.06146`. This approach introduced the concept of **transfer learning** to **natural language processing** (**NLP**).

Transfer learning is taking a model that has been trained on a large, general-purpose dataset and making it applicable to a specific use case by fine-tuning the large model with a smaller dataset that is specific to the use case.

The ULMFiT approach to transfer learning for NLP can be summarized as follows:

1. Start with a large language model trained on a large text dataset.

2. Fine-tune this language model on a text dataset that is related to a specific use case.

3. Use the fine-tuned language model to create a text classifier for the specific use case.

In this chapter, the large model is referred to as `AWD_LSTM`, and it has been trained on a big corpus taken from Wikipedia articles. We will fine-tune this large language model on datasets for two specific use cases: IMDb for movie reviews, and Covid tweets for social media posts regarding the Covid 19 pandemic. We then use each of the resulting fine-tuned language models to train text classifiers for each use case.

Here are the recipes that will be covered in this chapter:

*   Training a deep learning language model with a curated IMDb text dataset

*   Training a deep learning classification model with a curated text dataset

*   Training a deep learning language model with a standalone text dataset

*   Training a deep learning text classifier with a standalone text dataset

*   Test your knowledge

Technical requirements

Ensure that you have completed the setup sections from *Chapter 1*, *Getting Started with fastai*, and have a working **Gradient** instance or **Colab** setup. The recipes described in this chapter assume that you are using Gradient. Ensure that you have cloned the repository for the book from `https://github.com/PacktPublishing/Deep-Learning-with-fastai-Cookbook` and have access to the `ch4` folder. This folder contains the code samples described in this chapter.

Some of the examples in this chapter will take over an hour to run.

> **Note**
> Do not use Colab for these examples. With Colab you cannot control the GPU
> that you get for a session and these examples may run for many hours. For the
> examples in this chapter, use a for-pay GPU-enabled Gradient environment
> to ensure they complete in a reasonable time.

Training a deep learning language model with a curated IMDb text dataset

In this section, you will go through the process of training a language model on a curated text dataset using fastai. We take a pre-existing language model that is packaged with fastai and fine-tune it with one of the curated text datasets, IMDb, that contains text samples for the movie review use case. The result will be a language model with the broad language capability of the pre-existing language model, along with the use case-specific details of the IMDb dataset. This recipe illustrates one of the breakthroughs made by the team that created fastai, that is, transfer learning applied to NLP.

Getting ready

For the recipes so far in this book, we have recommended using the Gradient environment. You can use Gradient for this recipe and the instructions below include several workarounds to make the recipe work on Gradient. In particular, the pre-trained `AWD_LSTM` model will not be available in Gradient if its initial setup gets interrupted and the directory for the `IMDB` model is not writeable. If the setup of the pre-trained `AWD_LSTM` model gets interrupted, follow these steps:

1. In Colab, run the cells of the `text_model_training.ipynb` notebook up to and including the learner definition and training cell. Once you have done so, copy the contents of the `/root/.fastai/models/wt103-fwd` directory to a folder in your Drive environment.

2. Upload the files you copied in the previous step to the `/storage/models/wt103-fwd` directory in your Gradient environment.

With these steps, you should now be able to run the notebook for this recipe (and other recipes that make use of `AWD_LSTM`) in Gradient.

I am grateful for the opportunity to include the IMDB dataset featured in this chapter.

Dataset citation

Andrew L. Maas, Raymond E. Daly, Peter T. Pham, Dan Huang, Andrew Y. Ng, Christopher Potts. (2011) *Learning Word Vectors for Sentiment Analysis* (`https://ai.stanford.edu/~amaas/papers/wvSent_ acl2011.pdf`)

How to do it...

In this section, you will be running through the `text_model_training.ipynb` notebook to train a language model using the IMDb curated dataset. Once you have the notebook open in Colab, complete the following steps:

1. Run the cells in the notebook up to the `Training a language model` cell.

2. Run the following cell to define a `path` object associated with the IMDb curated dataset:

   ```
   path = untar_data(URLs.IMDB)
   ```

3. You can see the directory structure for this dataset in the output of the `tree -d` command run in the `imdb` directory (`/storage/data/imdb` in Gradient). Note that the labels for the dataset (whether a review is positive or negative) are encoded by the directory in which the text sample is located. For example, the negative review text samples in the training dataset are contained in the `train/ neg` directory.

   The following is the output of the `tree -d` command run in the `imdb` directory:

   ```
   ├── test
   │   ├── neg
   │   └── pos
   ├── tmp_clas
   ├── tmp_lm
   ├── train
   │   ├── neg
   │   └── pos
   └── unsup
   ```

4. Run the following cell to define a `TextDataLoaders` object:

```
dls =TextDataLoaders.from_folder(\
    path, valid = 'test', is_lm=True, bs=16)
```

Here are the arguments for the definition of the `TextDataLoaders` object:

a) `path`: The `path` object (associated with the IMDb curated dataset) that you defined earlier in the notebook.

b) `valid`: Identifies the folder in the dataset's directory structure that will be used to assess the performance of the model: `imdb/test`.

c) `is_lm`: Set to `True` to indicate that this object will be used for a language model (as opposed to a text classifier).

d) `bs`: Specifies the batch size.

> **Note**
>
> When you are training a language model with a large dataset such as IMDb, adjusting the `bs` value to be lower than the default batch size of `64` will be essential for avoiding memory errors, and that is why it is set to `16` in this `TextDataLoaders` definition.

5. Run the following cell to show a couple of items from a sample batch:

```
dls.show_batch(max_n=2)
```

The `max_n` argument specifies the number of sample batch items to show.

Note the output of this cell. The `text` column shows the original text. The `text_` column shows the same text shifted one token ahead, that is, it starts one word after the original text and ends one word past the original text. Given a sample such as one of the entries in the `text` column, the language model will predict the next word, as shown in the `text_` column. We can see the output of `show_batch()` in the following screenshot:

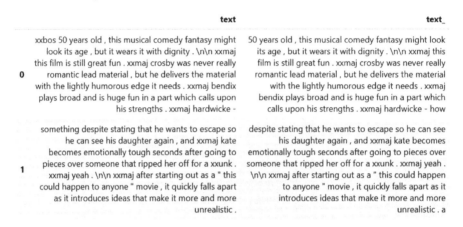

Figure 4.1 – Output of show_batch()

6. Run the following cell to define and train the deep learning model:

```
learn = language_model_learner(\
dls,AWD_LSTM, metrics=accuracy).to_fp16()
learn.fine_tune(1, 1e-2)
```

Here are the arguments for the definition of the `language_model_learner` object:

a) `dls`: The `TextDataLoaders` object that is defined previously in this notebook.

b) `AWD_LSTM`: The pre-trained model to use as a basis for this model. This is the pre-trained language model incorporated with fastai that is trained with Wikipedia. If you are running this notebook on Colab, you can find the files that make up this model in the `/root/.fastai/models/wt103-fwd` directory after you have run this cell.

c) `metrics`: The performance metric to be optimized for the model, in this case, accuracy.

Here are the arguments for the `fine_tune` statement:

a) The epoch number (first argument) specifies the number of epochs, that is, the number of times the algorithm goes through the full training data during the training process.

b) The learning rate (second argument) specifies the learning rate for the training process. The learning rate is the rate at which the algorithm moves toward learning optimal parameters.

> **Note**
>
> a) Depending on your environment, it may take more than an hour for this cell to run to completion. I strongly recommend that you use a for-pay GPU-enabled Gradient environment for this example and specify at least 3 hours for the instance to ensure that it completes in a reasonable time and that the instance doesn't shut down while this cell is running.
>
> b) The `language_model_learner` definition includes a call to `to_fp16()` to specify mixed-precision training (summarized here: `https://docs.fast.ai/callback.fp16.html#Learner.to_fp16`) to reduce the memory consumption of the training process and to prevent memory errors. Refer to the *There's more...* section for more details.

The output of the `fine_tune` statement shows the accuracy of the model and the time taken to complete the fine-tuning, as shown in the following screenshot:

epoch	train_loss	valid_loss	accuracy	time
0	4.741510	4.345975	0.266929	34:07

epoch	train_loss	valid_loss	accuracy	time
0	4.238236	4.044060	0.296340	35:18

```
CPU times: user 50min 5s, sys: 12min 56s, total: 1h 3min 2s
Wall time: 1h 9min 32s
```

Figure 4.2 – Output of the fine_tune statement

7. Run the following cell to exercise the language model you just trained:

```
learn.predict("what comes next", n_words=20)
```

Here are the arguments for this invocation of the language model:

a) The input text sample `"what comes next"` (first argument) is the phrase that the model will complete. The language model will predict what words should follow this phrase.

b) `n_words`: This is the number of words that the language model is supposed to predict to complete the input phrase.

The following screenshot shows an example of what the model's prediction could look like:

```
'what comes next is a same repugnantly western trash animation , badly written and watched TV comedy
, largely written by'
```

Figure 4.3 – The language model completes a phrase

8. Run the following cell to save the model. You can update the cell to specify the directory and filename to which to save the model:

```
learn.export('/notebooks/temp/models/lm_model_'+modifier)
```

9. Run the following cell to save the current path value:

```
keep_path = learn.path
```

10. Run the following cell to assign a new value to the learner object path. The reason for doing this is that the default location for the model is not writeable on Gradient so you need to change the path value to a directory where you have write access:

```
learn.path = Path('/notebooks/temp')
```

11. Run the following cell to save the encoder subset of the model. This is the model minus the final layer. You will use this in the *Training a deep learning classification model with a curated text dataset* section when you train a text classifier:

```
learn.save_encoder('ft_'+modifier)
```

Congratulations! You have successfully applied transfer learning to train a language model on the curated IMDb dataset. Note that the idea of applying transfer learning to NLP like this was only described for the first time in 2018. Now, thanks to the fastai framework, with just a few lines of code, you can take advantage of a technique that didn't exist scarcely 4 years ago!

How it works...

In this section, you have seen a simple example of how to train a language model with fastai deep learning using a curated text dataset. The language model is created by taking a model (`AWD_LSTM`) that has been pre-trained with the massive wiki dataset and then fine-tuning it using the IMDb dataset.

By taking advantage of transfer learning in this way, we end up with a language model that combines a good degree of capability on general-purpose English (thanks to the model pre-trained on the wiki dataset) as well as the capability to produce text that is specific to the use case of movie reviews (thanks to the IMDb dataset).

It's worthwhile to look a bit closer at the model in this recipe. A deeply detailed description of the model is beyond the scope of this book, so we will just focus on some highlights here.

As shown in the recipe, the model is defined as a `language_model_learner` object (documentation here: `https://docs.fast.ai/text.learner.html#language_model_learner`). This object is a specialization of the fastai `learner` object which you first saw in the *Understanding the world in four applications: tables, text, recommender systems* section, and in the images of *Chapter 1, Getting Started with fastai.*

The model in the recipe is based on the predefined `AWD_LSTM` model (documentation here: `https://docs.fast.ai/text.models.awdlstm.html#AWD_LSTM`). For this model, the output of `learn.summary()` shows only the high-level structure, including LSTM layers (fundamental to traditional NLP deep learning models) and dropout layers (used to reduce overfitting). Similarly, the output of `learn.model` for this model starts with layers for encoding (that is, transforming the input data to an intermediate representation used within the model) and ends with layers for decoding (that is, transforming the internal representation back to words).

There's more...

In this chapter, you will be working with some very large datasets, which means that you may need to take some extra steps to ensure that you don't run out of memory while you are preparing the datasets and training the model. Here, we'll describe some steps you can take to ensure that you train your fastai deep learning models on text datasets without running out of memory. We'll also go into more detail about how to save encoders in Gradient.

What happens if you run out of memory?

If you are trying to train a large model, you may get an out of memory message such as the following:

```
RuntimeError: CUDA out of memory. Tried to allocate 102.00 MiB
(GPU 0; 7.93 GiB total capacity; 7.14 GiB already allocated;
6.50 MiB free; 7.32 GiB reserved in total by PyTorch)
```

This message is saying that you have run out of memory on the GPU for your environment. What can you do if you encounter such a memory error? There are three steps you can take to get around this kind of memory error:

- Explicitly set the batch size.

- Use mixed-precision training.

- Ensure that you have only one notebook active at a time.

Memory error mitigation #1: Explicitly set the batch size

To explicitly set the batch size, you can restart the kernel for your notebook and then update the definition of the TextDataLoaders object to set the bs parameter, as shown here:

```
dls = TextDataLoaders.from_folder(untar_data(URLs.IMDB),
  valid='test', bs=16)
```

Setting the bs parameter explicitly specifies a batch size (the number of items on which the average loss is calculated) other than the default of 64. By explicitly setting the batch size to a smaller value than the default, you limit the amount of memory consumed by each training epoch (that is, a complete iteration through the training data). When you set the value of bs explicitly like this, ensure that the value you set for the bs parameter is a multiple of 8.

Memory error mitigation #2: Mixed-precision training

Another technique that you can use to control memory consumption is using **mixed-precision training**. You can specify mixed-precision training by applying the to_fp16() function to the definition of the learner object, as shown here:

```
learn = language_model_learner(dls,AWD_LSTM,
  drop_mult=0.5,metrics=accuracy).to_fp16()
```

By specifying this call to `to_fp16()`, you allow the model to be trained using floating-point numbers that are less precise and are therefore expressed with less memory. The result is that the model training process consumes less memory. Refer to the fastai documentation for more details: `https://docs.fast.ai/callback.fp16.html#Learner.to_fp16`.

Memory error mitigation #3: Stick to a single active notebook

Finally, another approach that you can take to prevent running out of memory is to run a single notebook at a time. On Gradient, for example, if you have multiple notebooks active at the same time, then you can exhaust your available memory. If you take the steps of setting a smaller batch size in the `TextDataLoaders` object and specifying `to_fp16()` in the learner object and still get memory errors, shut down the kernel of all the notebooks except the one you are currently working on.

In JuptyerLab in Gradient, you can shut down the kernel for a notebook by right-clicking on the notebook in the navigation pane and selecting **Shut Down Kernel** from the menu, as shown in the following screenshot:

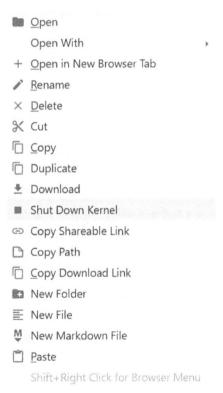

Figure 4.4 – Shutting down a kernel in Gradient JupyterLab

Workaround to allow you to save encoders

In addition to the memory tips we've just reviewed, there is one more tip you need to know for text models if you are using Gradient. On Gradient, you may run into a situation where you are not able to save and retrieve interim objects in the directory where fastai wants to save them.

For example, you need to save the encoder from the language model and then load that encoder when you train the text classifier. However, fastai forces you to save the encoder in the path for the dataset. The `save_encoder` function only lets you specify the unqualified file name, not the directory in which to save the encoder, as you can see in the following call to `save_encoder`:

```
learn.save_encoder('ft_'+modifier)
```

At the same time, in Gradient, the directory for the IMDb dataset, `/storage/data/imdb`, is read-only. So, how can you save an encoder if the directory where it must be saved is not writeable? You can work around this problem by temporarily updating the learner's `path` object, saving the encoder in the directory indicated by this temporary `path` value, and then setting the `path` object back to its original value, as shown here:

1. Save the path value for your model:

   ```
   keep_path = learner.path
   ```

2. Change the path value for your model to a directory that you have write access to, for example:

   ```
   learner.path = Path('/temp/models')
   ```

3. Save the model.

4. Change the path back to the original value:

   ```
   learner.path = keep_path
   ```

Training a deep learning classification model with a curated text dataset

In the previous section, we trained a language model using the curated text IMDb dataset. The model in the previous section predicted the next set of words that would follow a given set of words. In this section, we will take the language model that was fine-tuned on the IMDb dataset and use it to train a text classification model that classifies text samples that are specific to the movie review use case.

Getting ready

This recipe makes use of the encoder that you trained in the previous section, so ensure that you have followed the steps in the recipe in that section, in particular, that you have saved the encoder from the trained language model.

As mentioned in the previous section, you need to take some additional steps before you can run recipes in Gradient that use the language model pre-trained on the Wikipedia corpus. To ensure that you have access to the pre-trained language model that you need to use in this recipe, complete the following steps if the setup of AWD_LSTM was interrupted:

1. In Colab, run the cells of the `text_model_training.ipynb` notebook up to and including the learner definition and training cell. Once you have done so, copy the contents of the `/root/.fastai/models/wt103-fwd` directory to a folder in your Drive environment.

2. Upload the files you copied in the previous step to the `/storage/models/wt103-fwd` directory in your Gradient environment.

With these steps, you will be able to run the notebook for this recipe (and other recipes that make use of `AWD_LSTM`) in Gradient.

How to do it...

In this section, you will be running through the `text_classifier_model.ipynb` notebook to train a text classifier deep learning model using the `IMDb` curated dataset. Once you have the notebook open in Gradient, complete these steps:

1. Run the cells in the notebook up to the `Define the text classifier` cell.

2. Run the following cell to define a `TextDataLoaders` object:

```
dls_clas = TextDataLoaders.from_folder(\
path, valid='test')
```

Here are the arguments for the definition of the `TextDataLoaders` object:

a) `path`: Defines the path of the dataset used to define the `TextDataLoaders` object

b) `valid`: Identifies the folder in the dataset's directory structure that will be used to assess the performance of the model: `imdb/test`

3. Run the following cell to see a sample of entries from a batch:

```
dls_clas.show_batch(max_n=3)
```

4. The output of `show_batch()` shows text samples along with the class (indicated in the `category` column). fastai knows that the class is encoded in the directory where the text sample is and correctly renders it in `show_batch()`, as seen in the following screenshot:

	text	category
0	xxbos xxmaj match 1 : xxmaj tag xxmaj team xxmaj table xxmaj match xxmaj bubba xxmaj ray and xxmaj spike xxmaj dudley vs xxmaj eddie xxmaj guerrero and xxmaj chris xxmaj benoit xxmaj bubba xxmaj ray and xxmaj spike xxmaj dudley started things off with a xxmaj tag xxmaj team xxmaj table xxmaj match against xxmaj eddie xxmaj guerrero and xxmaj chris xxmaj benoit . xxmaj according to the rules of the match , both opponents have to go through tables in order to get the win . xxmaj benoit and xxmaj guerrero heated up early on by taking turns hammering first xxmaj spike and then xxmaj bubba xxmaj ray . a xxmaj german xxunk by xxmaj benoit to xxmaj bubba took the wind out of the xxmaj dudley brother . xxmaj spike tried to help his brother , but the referee restrained him while xxmaj benoit and xxmaj guerrero	pos
1	xxbos xxmaj by now you 've probably heard a bit about the new xxmaj disney dub of xxmaj miyazaki 's classic film , xxmaj laputa : xxmaj castle xxmaj in xxmaj the xxmaj sky . xxmaj during late summer of 1998 , xxmaj disney released " kiki 's xxmaj delivery xxmaj service " on video which included a preview of the xxmaj laputa dub saying it was due out in " 1 xxrep 3 9 " . xxmaj it 's obviously way past that year now , but the dub has been finally completed . xxmaj and it 's not " laputa : xxmaj castle xxmaj in xxmaj the xxmaj sky " , just " castle xxmaj in xxmaj sky " for the dub , since xxmaj laputa is not such a nice word in xxmaj spanish (even though they use the word xxmaj laputa many times	pos
2	xxbos xxmaj this movie was recently released on xxup dvd in the xxup us and i finally got the chance to see this hard - to - find gem . xxmaj it even came with original theatrical previews of other xxmaj italian horror classics like " xxunk " and " beyond xxup the xxup darkness " . xxmaj unfortunately , the previews were the best thing about this movie . \n\n " zombi 3 " in a bizarre way is actually linked to the infamous xxmaj lucio xxmaj fulci " zombie " franchise which began in 1979 . xxmaj similarly compared to " zombie " , " zombi 3 " consists of a threadbare plot and a handful of extremely bad actors that keeps this ' horror ' trash barely afloat . xxmaj the gore is nearly non - existent (unless one is frightened of people running around with	neg

Figure 4.5 – Output of show_batch()

5. Run the following cell to define the text classifier model:

```
learn_clas = text_classifier_learner(dls_clas, AWD_LSTM,
                                    metrics=accuracy).to_
fp16()
```

Here are the arguments for the definition of the `text_classifier_learner` object:

a) `dls_clas`: This is the `TextDataLoaders` object defined in the previous cell.

b) `AWD_LSTM`: This is the pre-trained model to use as a basis for this model. If you run this notebook in Colab, you can find the files that make up this model in the `/root/.fastai/models/wt103-fwd` directory after you have run this cell.

c) `metrics`: This is the performance metric to be optimized for the model, in this case, accuracy.

6. You need to load the encoder that you saved as part of the recipe in the previous section. The first step is to set the path for the `learn_clas` object so that it is the path in which the encoder is saved by running the following cell. Ensure that the directory specified is the directory where you saved the encoder in the previous recipe:

```
learn_clas.path = Path('/notebooks/temp')
```

7. Run the following cell to load the encoder that you saved in the recipe in the *Training a deep learning language model with a curated text dataset* section to the `learn_clas` object:

```
learn_clas = learn_clas.load_encoder('ft_'+modifier)
```

8. Run the following cell to train the model:

```
learn_clas.fit_one_cycle(5, 2e-2)
```

Here are the arguments for `fit_one_cycle`:

a) The argument epoch count (first argument) specifies that the training is run for 5 epochs.

b) The argument learning rate (second argument) specifies that the learning rate is equal to `0.02`.

The output of this cell shows the results of the training, including the accuracy and the time taken for each epoch, as shown in the following screenshot:

epoch	train_loss	valid_loss	accuracy	time
0	0.432951	0.299519	0.874200	03:46
1	0.411521	0.291931	0.878600	03:29
2	0.395382	0.275534	0.887120	03:28
3	0.409314	0.263998	0.890040	03:28
4	0.379574	0.264123	0.891080	03:28

Figure 4.6 – Results of training the text classification model

9. Run the cells to get predictions on text strings that you expect to be negative and positive and observe whether the trained model makes the expected predictions, as shown in the following screenshot:

```
# get a prediction on a negative phrase
preds = learn.predict("this film shows incredibly bad writing and is a complete disaster")

preds

('negative', TensorText(0), TensorText([0.7244, 0.2756]))

# get a prediction on a positive phrase
preds = learn.predict("this film shows incredible talent and is a complete triumph")

preds

('positive', TensorText(1), TensorText([0.0962, 0.9038]))
```

Figure 4.7 – Using the text classifier to get predictions on text strings

Congratulations! You have taken the language model that was fine-tuned on the IMDb dataset and used it to train a text classification model that classifies text samples that are specific to the movie review use case.

How it works...

You can see the power of fastai by contrasting the code in this section, which defines a text classifier, with the code in the previous section, which defines a language model. There are only three differences in total:

1. In the definition of the `TextDataLoaders` object, the following applies:

    a) The language model has the `is_lm` argument to indicate that the model is a language model.

    b) The text classifier has the `label_col` argument to indicate which column in the dataset contains the category that is being predicted by the model. In the case of the text classifier defined in this section, the label for the dataset is encoded in the directory structure of the dataset rather than as a column in the dataset, so this parameter is not needed in the definition of the `TextDataLoaders` object.

2. In the definition of the model, the following applies:

    a) The language model defines a `language_model_learner` object.

    b) The text classifier defines a `text_classifier_learner` object.

3. In getting a prediction from the model, the following applies:

    a) The language model takes two arguments for its call to `learn.predict()`, the string on which to make the prediction, and the number of words to predict.

    b) The text classifier takes one argument for its call to `learn.predict()`, the string whose class the model will predict.

With just these three differences, fastai takes care of all the underlying differences between a language model and a text classifier.

There's more...

If you are using the Gradient environment and you are using a notebook with a cost, you will want to control how long the notebook is active to avoid paying for more time than you need. You can select the duration of your session when you start it up by selecting an hour value from the **Auto-Shutdown** menu, as shown in the following screenshot:

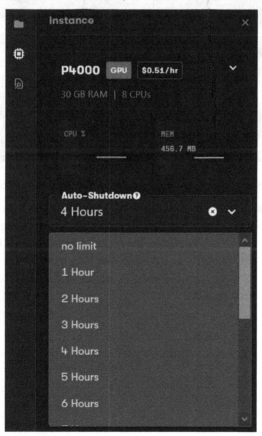

Figure 4.8 – Selecting a duration for a Gradient notebook session

Suppose that you end up selecting more time than you need and you are done with your session before the auto-shutdown time limit is reached. Should you explicitly shut down the session?

My experience has been that if you try to stop the instance in the Gradient notebook interface by selecting the **Stop Instance** button in the **Instance** view (as shown in *Figure 4.9*), you will risk putting your instance into a state where you cannot start it again easily:

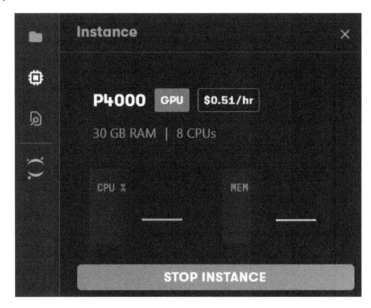

Figure 4.9 – Stop Instance button in the Instance view in Gradient

If you select **Stop Instance** and your instance gets into a state where you cannot start it again, then you will have to open a ticket with *Paperspace* support to fix your instance. After this happened to me a couple of times, I stopped using the **Stop Instance** button and just let the instance time out when I was finished working with it. You will save yourself time by never explicitly stopping your Gradient instance and instead just letting it time out when you are done with a session.

Training a deep learning language model with a standalone text dataset

In the previous sections, we trained a language model and a text classifier using the curated text dataset IMDb. In this section and the next section, we will train a language model and a text classifier using a standalone text dataset, the *Kaggle Coronavirus tweets NLP – Text Classification* dataset described here: `https://www.kaggle.com/datatattle/covid-19-nlp-text-classification`. This dataset includes a selection of tweets related to the Covid-19 pandemic, along with categorization for the tweets according to the following five categories:

- Extremely negative
- Negative
- Neutral
- Positive
- Extremely positive

The goal of the language model trained on this dataset is to predict the subsequent words in a Covid-related tweet given a starting phrase. The goal of the text classification model trained on this dataset, as described in the *Training a deep learning text classifier with a standalone text dataset* section, is to predict which of the five categories a phrase belongs in.

Getting ready

As mentioned in previous sections in this chapter, you need to take some additional steps before you can run this recipe in Gradient to ensure that you have access to the pre-trained language model that you will use in this recipe. If you have not done so already, follow these steps to prepare your Gradient environment if the setup of AWD_LSTM was interrupted:

1. In Colab, run the cells of the `text_model_training.ipynb` notebook up to and including the learner definition and training cell. Once you have done so, copy the contents of the `/root/.fastai/models/wt103-fwd` directory to a folder in your Drive environment.

2. Upload the files you copied in the previous step to the `/storage/models/wt103-fwd` directory in your Gradient environment.

With these steps, you should now be able to run the notebook for this recipe (and other recipes that make use of `AWD_LSTM`) in Gradient.

Ensure that you have uploaded the files that make up the standalone Covid-related tweets dataset to your Gradient environment by following these steps:

1. Download the `archive.zip` file from `https://www.kaggle.com/datatattle/covid-19-nlp-text-classification`.

2. Unzip the downloaded `archive.zip` file to extract the `Corona_NLP_test.csv` and `Corona_NLP_train.csv` files.

3. From the terminal in your Gradient environment, make `/storage/archive` your current directory:

   ```
   cd /storage/archive
   ```

4. Create the `/storage/archive/covid_tweets` directory:

   ```
   mkdir covid_tweets
   ```

5. Make `/storage/archive/covid_tweets` your current directory:

   ```
   cd /storage/archive/covid_tweets
   ```

6. Create `test` and `train` directories in `/storage/archive/covid_tweets`:

   ```
   mkdir test
   mkdir train
   ```

7. Upload the files you extracted in *step 2* (`Corona_NLP_test.csv` and `Corona_NLP_train.csv`) to `/storage/archive/covid_tweets`. You can use the upload button in JupyterLab in Gradient to do the upload, but you need to do it in several steps:

   a) From the terminal in your Gradient environment, make `/notebooks` your current directory:

   ```
   cd /notebooks
   ```

   b) If you have not already created a `notebooks/temp` directory, make a new `/notebooks/temp` directory:

   ```
   mkdir temp
   ```

c) In the JupyterLab file browser, make temp your current folder, select the upload button (see *Figure 4.10*), and select the Corona_NLP_test.csv and Corona_NLP_train.csv files from your local system folder where you extracted them in *step 2*:

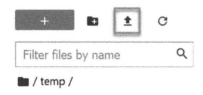

Figure 4.10 – Upload button in JupyterLab

d) From the terminal in your Gradient environment, copy the Corona_NLP_test.csv file into the /storage/archive/covid_tweets/test directory:

```
cp /notebooks/temp/Corona_NLP_test.csv /storage/archive/
covid_tweets/test/Corona_NLP_test.csv
```

e) Copy the Corona_NLP_train.csv file into the /storage/archive/covid_tweets/train directory:

```
cp /notebooks/temp/Corona_NLP_train.csv /storage/archive/
covid_tweets/train/Corona_NLP_train.csv
```

Once you have completed the steps to upload the files that make up the Covid-related tweets dataset, you should have the following directory structure in the /storage/archive/covid_tweets directory in your Gradient environment:

```
├── test
│      └── Corona_NLP_test.csv
└── train
       └── Corona_NLP_train.csv
```

With these preparation steps, you have brought the files that make up the dataset into the correct location in your Gradient environment to be used by a fastai model.

I am grateful for the opportunity to include the Covid tweets dataset in this book and I would like to thank the curators of this dataset and Kaggle for making the dataset available.

> **Dataset citation**
>
> Aman Miglani (2020). *Coronavirus tweets NLP - Text Classification*
> (`https://www.kaggle.com/datatattle/covid-19-nlp-text-classification`)

How to do it...

In this section, you will be running through the `text_standalone_dataset_lm.ipynb` notebook to train a language model using the Covid-related tweets standalone dataset. Once you have the notebook open in Gradient, complete these steps:

1. Run the cells in the notebook up to the `Ingest the dataset` cell.

2. Run the following cell to define a path object for the dataset:

   ```
   path = URLs.path('covid_tweets')
   ```

   > **Note**
   >
   > The argument for this path definition is the name of the root of the directory hierarchy in your Gradient environment into which you copied the CSV files for the dataset.

3. Run the following cell to define a `df_train` dataframe to contain the contents of the `Corona_NLP_train.csv` file:

   ```
   df_train = pd.read_csv(path/'train/Corona_NLP_train.csv',
                          encoding = "ISO-8859-1")
   ```

   Here are the arguments for the definition of the dataframe:

   a) The `path/'train/Corona_NLP_train.csv'` argument specifies the partially qualified filename for the training portion of the dataset.

   b) The `encoding = "ISO-8859-1"` argument specifies the encoding to use for the file. This encoding is selected to ensure that the content of the CSV file can be ingested into a dataframe without any errors.

4. Run the following cell to define a `TextDataLoaders` object:

```
dls = TextDataLoaders.from_df(df_train, path=path,
                              text_col='OriginalTweet',
                              is_lm=True)
```

Here are the arguments for the definition of the `TextDataLoaders` object:

a) `df_train`: The dataframe that you created in the previous step.

b) `path`: The path object for the dataset.

c) `text_col`: The column in the dataframe containing the text that will be used to train the model. For this dataset, the `OriginalTweet` column contains the text used to train the model.

d) `is_lm`: An indicator that this model is a language model.

5. Run the following cell to define and train the deep learning model with a `language_model_learner` object:

```
learn = language_model_learner(dls, AWD_LSTM,
                               metrics=accuracy).to_fp16()
learn.fine_tune(1, 1e-2)
```

> **Note**
>
> The definition of the `language_model_learner` object includes the call to `to_fp16()` to specify mixed-precision training (summarized here: `https://docs.fast.ai/callback.fp16.html#Learner.to_fp16`) to reduce the memory consumption of the training process.

Here are the arguments for the definition of the `language_model_learner` object:

a) `dls`: The `TextDataLoaders` object that you defined in the previous step.

b) `AWD_LSTM`: The pre-trained model to use as a basis for this model. This is the pre-trained language model incorporated with fastai that is trained with Wikipedia.

c) `metrics`: The performance metric to be optimized for the model, in this case, accuracy.

Here are the arguments for the `fine_tune` statement:

a) The epoch count argument (first argument) specifies the number of epochs for the training process.

b) The learning rate argument (second argument) specifies the learning rate for the training process.

The results of the training process, as shown in *Figure 4.11*, are displayed once the `fine_tune` statement has been run:

epoch	train_loss	valid_loss	accuracy	time
0	4.440168	3.958502	0.321858	01:56

epoch	train_loss	valid_loss	accuracy	time
0	3.998470	3.737445	0.343541	02:14

Figure 4.11 – Results of the training process

6. Run the following cell to exercise the trained language model:

```
learn.predict("what comes next", n_words=20)
```

The results are displayed as follows:

```
'what comes next to the message of panic for customers , who at this time constantly more hardship for bu
sinesses , they must'
```

Figure 4.12 – Prediction of a language model trained on a standalone text dataset

7. Run the following cell to save the model. You can update the cell to specify the directory and filename to which to save the model:

```
learn.export('/notebooks/temp/models/lm_model_
standalone'+modifier)
```

8. Run the following cell to save the current path value:

```
keep_path = learn.path
```

9. Run the following cell to assign a new value to the learner object path. The reason for doing this is that the default location for the model is not writeable on Gradient, so you need to change the path value to a directory where you have write access:

```
learn.path = Path('/notebooks/temp')
```

10. Run the following cell to save the encoder subset of the language model. This is the model minus the final layer. You will use this encoder in the next recipe when you train a text classifier on the Covid-related tweets standalone dataset:

```
learn.save_encoder('ft_standalone'+modifier)
```

Congratulations! You used fastai to do transfer learning on top of an existing model with the Covid-related tweets standalone dataset to create a language model fine-tuned on that dataset. In the next section, you will use the encoder that you saved in the last step to fine-tune a text classifier trained on the standalone dataset.

How it works...

In this section, you have seen a simple example of how to train a language model with fastai using a standalone text dataset. The language model is created by taking an existing model (AWD_LSTM) that has been trained with the massive Wikipedia dataset and then fine-tuning it using the standalone Covid-related tweets dataset.

By taking advantage of transfer learning in this way, we end up with a language model that combines a good degree of capability in terms of general-purpose English (thanks to the model pre-trained on the wiki dataset) as well as the capability to produce text that is specific to the use case of social media related to the Covid-19 pandemic (thanks to the Covid tweets dataset). By following the recipe in this section, you can take advantage of fastai to apply this approach (transfer learning for NLP) on other text datasets for other use cases.

Training a deep learning text classifier with a standalone text dataset

In the *Training a deep learning language model with a standalone text dataset* section, we trained a language model using the standalone text dataset: the Kaggle Coronavirus tweets NLP – Text Classification dataset described here: https://www.kaggle.com/datatattle/covid-19-nlp-text-classification. In this section, we will use this language model to create a text classifier trained with the Covid-related tweets dataset.

Getting ready

This recipe makes use of the encoder that you trained in the *Training a deep learning language model with a standalone text dataset* section, so ensure that you have followed the steps in the recipe in that section. In particular, ensure that you have saved the encoder from the language model you trained in the previous section.

Also, make sure you have followed all the steps from the *Getting ready* sub-section of the previous section to ensure the following:

- That you have access to the AWD_LSTM model in your Gradient environment
- That you have uploaded the files (Corona_NLP_test.csv and Corona_NLP_train.csv) that make up the standalone Covid-related tweets dataset to your Gradient environment

How to do it...

In this section, you will be running through the text_standalone_dataset_classifier.ipynb notebook to train a text classifier deep learning model using the Covid-related tweets dataset. Once you have the notebook open in Gradient, perform the following steps:

1. Run the cells in the notebook up to the Ingest the dataset cell.
2. Run the following cell to define a path object for the dataset. Note that the argument is the name of the root directory in your Gradient environment into which you copied the CSV files for the dataset:

   ```
   path = URLs.path('covid_tweets')
   ```

3. Run the following cell to define a dataframe to contain the contents of the Corona_NLP_train.csv file (the training portion of the Covid-related tweets dataset):

   ```
   df_train = pd.read_csv(path/'train/Corona_NLP_train.csv',
                      encoding = "ISO-8859-1")
   ```

   Here are the arguments for the definition of the dataframe:

   a) The path/'train/Corona_NLP_train.csv' argument specifies the partially qualified filename for the training portion of the dataset.

   b) The encoding = "ISO-8859-1" argument specifies the encoding to use for the file. This encoding is selected to ensure that the content of the CSV file can be ingested into a dataframe without any errors.

4. Run the following cell to define a `TextDataLoaders` object:

    ```
    dls = TextDataLoaders.from_df(df_train, path=path, text_
    col='OriginalTweet',label_col='Sentiment')
    ```

    Here are the arguments for the definition of the `TextDataLoaders` object:

    a) `df_train`: The dataframe that you created in the previous step.

    b) `path`: The path object for the dataset.

    c) `text_col`: The column in the dataframe containing the text that will be used to train the model. For this dataset, the `OriginalTweet` column contains the text used to train the model.

    d) `label_col`: The column in the dataframe containing the labels that the text classifier will predict.

5. Run the following cell to see a batch from the `TextDataLoaders` object that you defined in the previous step:

    ```
    dls.show_batch(max_n=3)
    ```

    The output of this statement, the `text` and `category` columns, will be as follows:

	text	category
0	xxbos xxrep 5 ? ? ? xxrep 7 ? ? ? xxrep 7 ? xxrep 4 ? xxrep 4 ? xxrep 11 ? ? ? xxrep 6 ? xxrep 4 ? , xxrep 3 ? xxrep 3 ? ? ? xxrep 3 ? xxrep 4 ? xxrep 3 ? ? ? ? xxrep 4 ? ? ? xxrep 3 ? , xxrep 4 ? ? ? ? ? xxrep 6 ? xxrep 3 ? xxrep 3 ? ? ? xxrep 3 ? \r\r\n__ xxrep 5 ? xxrep 6 ? ? ? xxrep 3 ? xxrep 4 ? xxrep 4 ? ? ? xxrep 4 ? xxrep 6 ? xxrep 4 ? xxrep 8 ? ? ? xxrep 6 ? ? ? xxrep 5 ? ? ? xxrep 3 ? xxrep 4 ? ? ? xxrep 7 ? xxrep 5 ? - xxrep 8 ? xxrep 5	Neutral
1	xxbos xxmaj fun xxmaj riding 4 xxmaj xxunk , xxmaj shield xxmaj bash # xxmaj cod # callofduty # xxmaj practice # xxmaj xxunk # xxmaj xxunk # xxmaj recreation # xxmaj fun # xxmaj bored # todo # xxmaj coronavirus # xxmaj quarantine # xxmaj isolation # toiletpaper # xxmaj lockdown # xxmaj art # xxmaj milk # xxmaj water # xxmaj xxunk # xxmaj weather # xxmaj cleveland # xxmaj ohio # xxmaj browns # xxup nfl # xxmaj xxunk # xxmaj poetry \r\r\n https : / / t.co / xxunk via @youtube	Positive
2	xxbos xxmaj friends ! xxmaj it 's xxmaj march 25 , 2020 at 03:00pm- time to xxup stop xxup renting & & buy a # home from # realtor xxmaj kally (khoelcher (at) gmail (dot) com) of # xxmaj goodyear # xxmaj arizona # coldwellbanker (269)240 - 8824 . # xxup n95 masks , # gloves , & & hand # sanitizer provided to xxup prevent # coronavirus . # xxmaj avondale # xxmaj buckeye # å __ https : / / t.co / xxunk	Extremely Positive

Figure 4.13 – Batch from the Covid tweets dataset

6. Run the following cell to define the `text_classifier_learner` object for the text classifier model:

    ```
    learn_clas = text_classifier_learner(dls, AWD_LSTM,
                                metrics=accuracy).to_fp16()
    ```

> **Note**
>
> The definition of the `text_classifier_learner` object includes the call to `to_fp16()` to specify mixed-precision training (summarized here: `https://docs.fast.ai/callback.fp16.html#Learner.to_fp16`) to reduce the memory consumption of the training process.

Here are the arguments for the definition of the `text_classifier_learner` object:

a) `dls`: The `TextDataLoaders` object that you defined in the previous step.

b) `AWD_LSTM`: The pre-trained model to use as a basis for this model. This is the pre-trained language model incorporated with fastai that is trained with Wikipedia.

c) `metrics`: The performance metric to be optimized for the model, in this case, accuracy.

7. Run the following cell to assign a new value to the learner object path. The reason for doing this is to set the path to match the directory where you saved the encoder in the recipe in the previous section:

```
learn_clas.path = Path('/notebooks/temp')
```

8. Run the following cell to load the encoder that you saved in the recipe in the *Training a deep learning language model with a standalone text dataset* section to the `learn_clas` object:

```
learn_clas =\
learn_clas.load_encoder('ft_standalone'+modifier)
```

9. Run the following cell to reset the value of the learner object path:

```
learn_clas.path = keep_path
```

10. Run the following cell to train the model:

```
learn_clas.fit_one_cycle(1, 2e-2)
```

Here are the arguments for `fit_one_cycle`:

a) The epoch count argument (first argument) specifies that the training is run for 1 epoch.

b) The learning rate argument (second argument) specifies that the learning rate is equal to 0.02.

The output of this cell (as shown in *Figure 4.14*) shows the results of the training:

epoch	train_loss	valid_loss	accuracy	time
0	1.461614	1.281477	0.453286	00:37

Figure 4.14 – Output of the text classifier training

11. Run the cells to get predictions on text strings that you expect to be negative and positive and observe whether the trained model makes the expected predictions (as shown in *Figure 4.15*):

```
preds = learn_clas.predict("the government's approach to the pandemic has been a complete disaster")
```

```
preds
```

```
('Negative',
 TensorText(2),
 TensorText([0.3328, 0.0545, 0.3551, 0.1026, 0.1551]))
```

```
preds = learn_clas.predict("the new vaccines hold the promise of a quick return to economic growth")
```

```
preds
```

```
('Extremely Positive',
 TensorText(1),
 TensorText([0.0565, 0.3758, 0.1528, 0.0699, 0.3450]))
```

Figure 4.15 – Using the text classifier to get predictions on text strings

Congratulations! You have taken advantage of the facilities of fastai to train a text classifier on a standalone dataset using transfer learning.

How it works...

The code for the text classifier model for the standalone Covid-related tweets dataset has some differences from the code for the text classifier model for the curated IMDb text dataset. Let's examine some of these differences.

For the IMDb dataset, the `TextDataLoaders` definition does not include a `label_col` parameter:

```
dls_clas = TextDataLoaders.from_folder(path, valid='test')
```

By contrast, the `TextDataLoaders` definition for the standalone dataset includes both `text_col` and `label_col` parameters:

```
dls = TextDataLoaders.from_df(df_train, path=path, text_
col='OriginalTweet',label_col='Sentiment')
```

What's the reason for these differences? First, for the `IMDb` dataset, we use the `from_folder` variation of `TextDataLoaders` because the dataset is organized as a collection of individual text files whose class is encoded by the directory that the file is in. Here is the directory structure of the IMDb dataset:

```
├── test
│   ├── neg
│   └── pos
├── tmp_clas
├── tmp_lm
├── train
│   ├── neg
│   └── pos
└── unsup
```

Consider one file from the IMDB dataset, `train/pos/9971_10.txt`:

This film was Excellent, I thought that the original one was quiet mediocre. This one however got all the ingredients, a factory 1970 Hemi Challenger with 4 speed transmission that really shows that Mother Mopar knew how to build the best muscle cars! I was in Chrysler heaven every time Kowalski floored that big block Hemi, and he sure did that a lot :)

How does fastai know the class of this review when we train the text classifier model? It knows because this file is in the `/pos` directory. Thanks to the flexibility of fastai, we simply have to pass the `path` value to the definition of the `TextDataLoaders` object and the fastai framework figures out the category of each text sample in the dataset.

Now, let's consider the standalone Covid-related tweets dataset. This dataset is packaged as CSV files that look like this:

UserName	ScreenName	Location	TweetAt	OriginalTweet	Sentiment
3799	48751	London	16-03-2020	@MeNyrbie @Phil_Gahan @Chrisitv https://t.co/iFz9FAn2Pa and htt	Neutral
3800	48752	UK	16-03-2020	advice Talk to your neighbours family to exchange phone numbers ci	Positive
3801	48753	Vagabonds	16-03-2020	Coronavirus Australia: Woolworths to give elderly, disabled dedicate	Positive
3802	48754		16-03-2020	My food stock is not the only one which is empty...	Positive

Figure 4.16 – Sample from the Covid-related tweets dataset

Unlike the IMDb dataset, the Covid-related tweets dataset is encapsulated in just two files (one for the training dataset and one for the test dataset). The columns of these files have the information that fastai needs to train the model:

- The text of the sample – in the `OriginalTweet` column
- The class (also known as the label) of the sample – in the `Sentiment` column

To tell fastai how to interpret this dataset, we need to explicitly tell it which column of the dataset the text is in and which column the class (or label) is in the definition of the `TextDataLoaders` object, as follows:

```
dls = TextDataLoaders.from_df(df_train, path=path, text_
col='OriginalTweet',label_col='Sentiment')
```

The `IMDb` dataset is made up of thousands of individual text files spread across a complex set of directories that encode the class of each text file. By contrast, the Covid-related tweets dataset is made up of two CSV files that have the text samples and their classes as columns. Despite the differences in the organization of these two datasets, fastai can ingest them and prepare them to train a deep learning model with just a few tweaks to the definition of the `TextDataLoaders` object. fastai's ability to easily ingest datasets in a variety of different formats isn't just useful for text datasets; it is useful for all kinds of datasets. As you will see in *Chapter 6, Training Models with Visual Data*, we really benefit from this ability when we deal with image datasets, which have many different kinds of organization.

Test your knowledge

Now that you have worked through a number of extended examples of training fastai deep learning models with text datasets, you can try some variations to practice what you've learned.

Getting ready

Ensure that you have followed the *Getting ready* steps from the *Training a deep learning text classifier with a standalone text dataset* section to prepare your Gradient environment and upload the Covid-related tweets dataset.

How to do it...

You can follow the steps in this section to try some variations on the models that you trained with the Covid-related tweets dataset:

1. Make a copy of the `text_standalone_dataset_lm.ipynb` notebook that you worked through in the *Training a deep learning language model with a standalone text dataset* recipe. Give your new copy of the notebook the following name: `text_standalone_dataset_lm_combo.ipynb`.

2. In your new notebook, in addition to creating a dataframe for the train CSV `Corona_NLP_train.csv` file, create a dataframe for the test CSV `Corona_NLP_test.csv` file by adding a cell to the notebook that looks like this:

   ```
   df_test = pd.read_csv(path/'test/Corona_NLP_test.csv
   ',encoding = "ISO-8859-1")
   ```

3. Use the pandas `concat` function to combine the two dataframes into a new dataframe called `df_combo`:

   ```
   df_combo = pd.concat([df_train, df_test], axis=0)
   ```

4. Now update the remainder of your new notebook to use `df_combo` instead of `df_train` and run the whole notebook to train a new language model. Do you notice any difference in the performance of the model?

5. In most model training situations, you need to ensure that you don't use the test dataset to train the model. Can you think of why you could get away with using the test set to train a language model like this?

Congratulations! You have completed your review of training fastai deep learning models on text datasets using fastai.

5
Training Recommender Systems

In this book, so far we have worked through recipes to train deep learning with fastai for a variety of datasets. In this chapter, we will go through recipes that take advantage of fastai's support for **recommender systems**, also known as **collaborative filtering systems**. Recommender systems combine the characteristics of tabular data models introduced in *Chapter 3, Training Models with Tabular Data*, with characteristics of text data models introduced in *Chapter 4, Training Models with Text Data*.

Recommender systems cover a narrow, but well-established, use case: given a set of users and their ratings of a set of items, a recommender system predicts the rating that a user will give for an item that the user has not rated yet. For example, given a set of books and a set of readers' assessments of these books, recommender systems can make predictions about a given reader's assessment of a book they haven't read yet.

In this chapter, you will learn how to use fastai's built-in support for recommender systems by working through a series of recipes that train models on a variety of recommender system datasets. You will see fastai features that will be familiar to you from previous chapters, as well as some new features that are unique to recommender systems. By the time you have completed this chapter, you will be ready to use the fastai high-level API to create recommender systems on your own datasets.

Here are the recipes that will be covered in this chapter:

- Training a recommender system on a small curated dataset
- Training a recommender system on a large curated dataset
- Training a recommender system on a standalone dataset
- Test your knowledge

Technical requirements

Ensure that you have completed the setup sections from *Chapter 1*, *Getting Started with fastai*, and have a working **Gradient** instance or **Colab** setup. The recipes described in this chapter assume that you are using Gradient. Ensure that you have cloned the repository for the book, `https://github.com/PacktPublishing/Deep-Learning-with-fastai-Cookbook`, and have access to the ch5 folder. This folder contains the code samples described in this chapter.

Training a recommender system on a small curated dataset

You may recall that *Chapter 1*, *Getting Started with fastai*, described applications supported by fastai to cover four types of datasets: **tabular**, **text**, **recommender systems**, and **images**. In *Chapter 2*, *Exploring and Cleaning Up Data with fastai*, you saw sections on examining tabular datasets, text datasets, and image datasets.

You may have wondered why there wasn't a section on examining recommender system datasets. The reason is that the data ingestion process for recommender systems in fastai is identical to the process for tabular datasets, as you will see in this section. While the ingestion process for recommender systems in fastai is identical to the ingestion process for tabular datasets, fastai does provide model training details that are specifically intended for recommender systems.

In this section, we will go through the process of training a recommender system on a curated dataset to learn how to train recommender systems with fastai.

In this section, you will train your recommender system using a small curated dataset – ML_SAMPLE. This dataset is a subset of the MovieLens dataset (https://grouplens. org/datasets/movielens) that contains user scores for movies. In the *Training a recommender system on a large curated dataset* section, we will train a recommender system on a much larger subset of the MovieLens dataset – ML_100k.

Getting ready

Confirm that you can open the training_recommender_systems.ipynb notebook in the ch5 directory of your repository.

The dataset used in this section and *Training a recommender system on a large curated dataset* section are from the MovieLens Datasets. I gratefully acknowledge the opportunity to use this dataset to illustrate the recommender system capabilities of fastai.

> **Dataset citation**
>
> Andrew L. Maas, Raymond E. Daly, Peter T. Pham, Dan Huang, Andrew Y. Ng, and Christopher Potts. (2011). *Learning Word Vectors for Sentiment Analysis* (http://ai.stanford.edu/~amaas/papers/ wvSent_acl2011.pdf). The 49th Annual Meeting of the Association for Computational Linguistics (ACL 2011) http://www.aclweb.org/ anthology/P11-1015

How to do it...

In this section, you will be running through the training_recommender_systems. ipynb notebook. Once you have the notebook open in your fastai environment, complete the following steps:

1. Run the cells in the notebook up to the Ingest the dataset cell to import the required libraries and set up your notebook.

2. Run the following cell to define the path object for this dataset:

    ```
    path = untar_data(URLs.ML_SAMPLE)
    ```

3. Run the following cell to examine the directory structure of the dataset:

    ```
    path.ls()
    ```

The following output shows the directory structure of the dataset:

```
(#1) [Path('/storage/data/movie_lens_sample/ratings.csv')]
```

Figure 5.1 – Output of path.ls()

4. Run the following cell to load the dataset into the df DataFrame:

```
df = pd.read_csv(path/'ratings.csv')
```

5. Run the following cell to see some records from the dataset:

```
df.head()
```

The output, as shown in *Figure 5.2*, lists records from the dataset. Each record represents a user's rating for a movie. The userId column has the ID for the user. The movieId column has the ID for the movie. The values in the rating column are the ratings given by the users for the movies. For example, user 73 gives movie 1097 a rating of 4.0:

	userId	movieId	rating	timestamp
0	73	1097	4.0	1255504951
1	561	924	3.5	1172695223
2	157	260	3.5	1291598691
3	358	1210	5.0	957481884
4	130	316	2.0	1138999234

Figure 5.2 – Output of df.head()

6. Run the following cell to define a CollabDataLoaders object for the dataset:

```
dls=CollabDataLoaders.from_df(df,bs= 64)
```

The definition of the CollabDataLoaders object uses the following arguments:

a) df: The DataFrame that you defined earlier in this notebook

b) bs: The batch size for the model training process

7. Run the following cell to see a batch from the `CollabDataLoaders` object that you defined in the previous step:

```
dls.show_batch()
```

The output displays the contents of the batch, as shown in the following screenshot:

	userId	movieId	rating
0	461	150	3.0
1	285	3793	4.0
2	15	377	4.0
3	468	47	3.5
4	587	318	4.5
5	285	1291	3.0
6	268	500	3.0
7	380	590	4.0
8	128	364	5.0
9	574	780	3.5

Figure 5.3 – Output of show_batch()

8. Run the following cell to define the `collab_learner` object for the recommender system model:

```
learn=collab_learner(dls, y_range= [ 0 , 5.0 ] )
```

Here are the arguments for the definition of the `collab_learner` object:

a) `dls`: The `CollabDataLoaders` object that you defined in a previous step

b) `y_range`: The range of values in the `rating` column

9. Run the following cell to train the collaborative filtering model:

```
learn.fit_one_cycle( 5 )
```

Here is the argument for the model definition:

a) 5: The number of epochs in the training run for the model

The output shows the training and validation loss, as shown in the following screenshot. You get an indication of the training improving as the validation loss decreases through the epochs:

epoch	train_loss	valid_loss	time
0	2.568535	2.594522	00:00
1	2.278521	2.051913	00:00
2	1.661850	1.358955	00:00
3	1.235316	1.137351	00:00
4	1.060143	1.107881	00:00

Figure 5.4 – Output of training the collaborative filtering model

10. Now that we have trained a recommender system, we want to test it out by getting it to predict the ratings that some users will give for some movies. To exercise the recommender system, we first have to create some test data. To create some test data, run the following cell to define a DataFrame that includes a couple of test entries:

```
scoring_columns = ['userId','movieId']
test_df = pd.DataFrame(columns=scoring_columns)
test_df.at[0,'userId'] = 388
test_df.at[0,'movieId'] = 153
test_df.at[1,'userId'] = 607
test_df.at[1,'movieId'] = 1210
test_df.head()
```

Here are the key elements of this cell:

a) scoring_columns: A list of the column names of the DataFrame. Note that these column names are the same column names that you saw in the output of show_batch().

b) test_df: The DataFrame for containing the test entries with column names specified in the scoring_columns list.

The output of `test_df.head()` shows the contents of the completed DataFrame, as shown in the following screenshot:

	userId	movieId
0	388	153
1	607	1210

Figure 5.5 – Contents of the test_df DataFrame

What do these entries in the `test_df` DataFrame mean? We will use this DataFrame to see what the recommender system will predict for the following combinations:

a) The rating that `userId 388` will give for `movieId 153`.

b) The rating that `userId 607` will give for `movieId 1210`.

11. Now that you have defined the `test_df` DataFrame to contain the entries you want to test the recommender system with, you can use it to exercise the trained recommender system. Run the following cell to get rating predictions from the recommender system for the entries in `test_df`:

```
dl = learn.dls.test_dl(test_df)
learn.get_preds(dl=dl)
```

The output of this cell shows the results of the recommender system, as shown in the following screenshot:

```
(tensor([2.4156, 3.6090]), None)
```

Figure 5.6 – Results of the recommender system

What do these results mean? The recommender system is predicting the following:

a) `userId 388` will give a rating of `2.4156` for `movieId 153`.

b) `userId 607` will give a rating of `3.6090` for `movieId 1210`.

Congratulations! You have trained and exercised a recommender system in fastai using one of the curated datasets.

How it works...

In this section, you worked through a recipe to train a very basic recommender system. Here's a summary of the key fastai objects you created in this recipe:

- A `path` object associated with the `URLs` object for the curated dataset:

  ```
  path = untar_data(URLs.ML_SAMPLE)
  ```

- A `DataLoaders` object:

  ```
  dls=CollabDataLoaders.from_df(df,bs= 64)
  ```

- A `learner` object:

  ```
  learn=collab_learner(dls, y_range= [ 0 , 5.0 ]
  ```

These ingredients should look familiar to you; they are the same core ingredients that you used in the recipes in *Chapter 3*, *Training Models with Tabular Data*, and *Chapter 4*, *Training Models with Text Data*. One of the strengths of the high-level fastai API is that it uses the same building blocks to create and train deep learning models for a broad variety of datasets, including the datasets you have seen in previous chapters and in this section.

There is one part of this section that is different from most of the recipes that you have seen so far – exercising the trained deep learning model on a set of test samples. Recall how you exercised the language model in *Chapter 4*, *Training Models with Text Data*. Here is the cell from the `text_model_training.ipynb` notebook that exercises the language model:

```
learn.predict("what comes next", n_words=20)
```

With a simple call to the `predict()` function, you get the results from the trained language model, as shown in the following screenshot:

```
'what comes next to the message of panic for customers , who at this time constantly more hardship for bu
sinesses , they must'
```

Figure 5.7 – Prediction of a language model trained on a standalone text dataset

We will contrast the way that you exercised the language model in *Chapter 4*, *Training Models with Text Data*, with what you needed to do in this recipe to train a recommender system as follows:

1. Build a `test_df` DataFrame to contain the test samples, as shown in the following screenshot:

	userId	movieId
0	388	153
1	607	1210

Figure 5.8 – Contents of the test_df DataFrame

2. Call test_dl using the test_df DataFrame that you just created as an argument. You can think of test_dl applying the same pipeline to the test_df DataFrame that was applied to the CollabDataLoaders object, dls:

```
dl = learn.dls.test_dl(test_df)
```

The output of this call is the test data loader object, dl. With this call to test_dl, you have transformed the DataFrame into a format to which the trained model (the recommender system) can be applied.

3. Call get_preds on the learn object to apply the trained recommender system to the data loader object, dl:

```
learn.get_preds(dl=dl)
```

Why did you need the extra steps to exercise the recommender system that you didn't need when you were exercising the language model? The answer is that the text phrase that is the input to the language model does not need to go through a sample-specific pipeline before it can be applied to the model.

Behind the scenes, fastai uses the vocabulary associated with the TextDataLoaders object to convert the input string into tokens (by default, words) and then convert the tokens into numeric IDs. You don't have to explicitly invoke this pipeline – it gets invoked implicitly for you when you call predict() on the learn object for the language model.

Unlike the text string that feeds into the language model, the input to the recommender system has structure, and that's why you need to define the input sample for the recommender system as a DataFrame. Once you have defined the DataFrame, you need to call test_dl to apply the transformation pipeline (that was implicitly defined when you defined the CollabDataLoaders object) on the input sample. The output of test_dl can then be used to get a prediction from the recommender system.

Before moving on to the next recipe, it's worth taking a closer look at the model from this recipe. A deeply detailed description of the model is beyond the scope of this book, so we will just focus on some highlights here.

As shown in the recipe, the model is defined as a `collab_learner` object (documentation here: `https://docs.fast.ai/collab.html#collab_learner`). This object is a specialization of the fastai `learner` object which you first saw in section *Understanding the world in four applications: tables, text, recommender systems, and images* of *Chapter 1, Getting Started with fastai*.

The output of `learn.model` shows that this model is an `EmbeddingDotBias` model (documentation here: `https://docs.fast.ai/collab.html#EmbeddingDotBias`).

Training a recommender system on a large curated dataset

In the *Training a recommender system on a small curated dataset* section, we saw the basics of how to create a recommender system model. The resulting system left something to be desired because the dataset only included user IDs and movie IDs, so it wasn't possible to determine what movies were actually being rated by users and having their ratings predicted by the model.

In this section, we are going to create a recommender system that addresses this gap in the previous recommender system because it is trained on a dataset that includes movie titles. Like the `ML_SAMPLE` dataset, the dataset we'll use in this section, `ML_100k`, is also derived from the MovieLens dataset, but it includes a much larger set of records and a much richer set of features. By creating a recommender system using this dataset, we will encounter additional features in fastai for ingesting and working with recommender system datasets and get a trained recommender system that is more interesting to use.

Getting ready

Confirm that you can open the `training_large_recommender_systems.ipynb` notebook in the `ch5` directory of your repository.

In the recipe in this section, you will use the `tree` command to examine the directory that contains the dataset. If you have not already installed the `tree` command in your Gradient instance, follow these steps to install it:

1. Run the following command in a Gradient terminal to prepare to install the `tree` command:

    ```
    apt update
    ```

2. Run the following command in a Gradient terminal to install the `tree` command:

```
apt install tree
```

Now that you have the `tree` command available to use in your Gradient environment, you are all ready to work through the recipe in this section.

How to do it...

In this section, you will be running through the `training_large_recommender_systems.ipynb` notebook. Once you have the notebook open in your fastai environment, complete the following steps:

1. Run the cells in the notebook up to the `Ingest the dataset` cell to import the required libraries and set up your notebook.

2. Run the following cell to define the `path` object for this dataset:

```
path = untar_data(URLs.ML_100k)
```

3. Run the following cell to examine the directory structure of the dataset:

```
path.ls()
```

The output shows the directory structure of the dataset, as shown in *Figure 5.9*. Note that the dataset has a more complex structure than `ML_SAMPLE`:

```
(#23) [Path('/storage/data/ml-100k/ua.base'),Path('/storage/data/ml-100k/README'),Path('/storage/data/ml-100k/u4.test'),Pat
h('/storage/data/ml-100k/u.genre'),Path('/storage/data/ml-100k/u.item'),Path('/storage/data/ml-100k/u2.test'),Path('/storag
e/data/ml-100k/u.user'),Path('/storage/data/ml-100k/u5.test'),Path('/storage/data/ml-100k/u.occupation'),Path('/storage/dat
a/ml-100k/u5.base')...]
```

Figure 5.9 – Output of path.ls() for the ML_100k dataset

4. You can get an idea of the files that make up the dataset by examining the path directly. You can do this by running the `tree` command. First, run the following command in a Gradient terminal to make the root directory of the dataset your current directory:

```
cd /storage/data/ml-100k
```

5. Next, run the following command in the Gradient terminal to list the contents of the directory:

```
tree
```

The output of the `tree` command lists the contents of the directory:

```
├── README
├── allbut.pl
├── mku.sh
├── u.data
├── u.genre
├── u.info
├── u.item
├── u.occupation
├── u.user
├── u1.base
├── u1.test
├── u2.base
├── u2.test
├── u3.base
├── u3.test
├── u4.base
├── u4.test
├── u5.base
├── u5.test
├── ua.base
├── ua.test
├── ub.base
└── ub.test
```

We want to focus on the contents of two files from this set: `u.data`, which lists the ratings provided by users for movie IDs, and `u.item`, which lists details about the movies, including their titles.

6. Run the following cell to define the `df_data` DataFrame to contain the contents of the `u.data` file. This DataFrame will contain the ratings provided by users for movies:

```
df_data = pd.read_csv(path/'u.data', delimiter = '\t',
header = None, \
names = ['userId','movieId','rating','timestamp'])
```

This DataFrame definition is the most complex one that we have seen so far. Let's go through the arguments:

a) `path/'u.data'`: Specifies the source for this DataFrame, the `u.data` file

b) `delimiter = '\t'`: Specifies that tabs are the delimiter that separates columns in this file

c) `header = None`: Specifies that the `u.data` file does not include column names in the first row

d) `names = ['userId','movieId','rating','timestamp']`: Assigns names to the columns of the DataFrame

7. Run the following cell to define the `df_item` DataFrame to contain the contents of the `u.item` file. This DataFrame will contain details about movies, including their titles:

```
df_item = pd.read_csv(path/'u.item', delimiter =
'|',header = None,encoding = "ISO-8859-1")
```

Here are the arguments for the definition of `df_item`:

a) `path/'u.item'`: Specifies the source for this DataFrame, the `u.item` file

b) `delimiter = '|'`: Specifies that the pipe character, '|', is the delimiter that separates columns in this file

c) `header = None`: Specifies that this file does not include column names in the first row

d) `encoding = "ISO-8859-1"`: Specifies the encoding used to read the file. You will get an error if you do not specify this encoding

8. Run the following cell to see a sample of the contents of the `df_data` DataFrame that you created:

```
df_data.head()
```

The output of this command shows the first few rows of the DataFrame:

	userId	movieId	rating	timestamp
0	196	242	3	881250949
1	186	302	3	891717742
2	22	377	1	878887116
3	244	51	2	880606923
4	166	346	1	886397596

Figure 5.10 – The first few rows of the df_data DataFrame

9. Run the following cell to see the dimensions of the `df_data` DataFrame:

```
df_data.shape
```

The output of this command shows the number of rows and columns in the df_data DataFrame:

$$(100000, 4)$$

Figure 5.11 – Shape of the df_data DataFrame

10. Run the following cell to see a sample of the contents of the `df_item` DataFrame that you created:

```
df_item.head()
```

The output of this command shows the first few rows of the DataFrame:

	0	1	2	3	4	5	6	7	8	9	10	11	12	13	14	15	16	17	18	19	20	21	22	23
0	1	Toy Story (1995)	01-Jan-1995	NaN	http://us.imdb.com/M/title-exact?Toy%20Story%20(1995)	0	0	0	1	1	1	0	0	0	0	0	0	0	0	0	0	0	0	0
1	2	GoldenEye (1995)	01-Jan-1995	NaN	http://us.imdb.com/M/title-exact?GoldenEye%20(1995)	0	1	1	0	0	0	0	0	0	0	0	0	0	0	0	0	1	0	0
2	3	Four Rooms (1995)	01-Jan-1995	NaN	http://us.imdb.com/M/title-exact?Four%20Rooms%20(1995)	0	0	0	0	0	0	0	0	0	0	0	0	0	0	0	0	1	0	0
3	4	Get Shorty (1995)	01-Jan-1995	NaN	http://us.imdb.com/M/title-exact?Get%20Shorty%20(1995)	0	1	0	0	0	1	0	0	1	0	0	0	0	0	0	0	0	0	0
4	5	Copycat (1995)	01-Jan-1995	NaN	http://us.imdb.com/M/title-exact?Copycat%20(1995)	0	0	0	0	0	0	1	0	1	0	0	0	0	0	0	1	0	0

Figure 5.12 – The first few rows of the df_item DataFrame

11. Run the following cell to see the dimensions of the `df_item` DataFrame:

```
df_item.shape
```

The output of this command shows the number of rows and columns in the df_
item DataFrame:

$$(1682, 24)$$

Figure 5.13 – The shape of the df_item DataFrame

12. Run the following cell to prepare the `df_item` DataFrame by removing most of the columns and adding column names for the remaining columns:

```
df_item = df_item.iloc[:,0:2]
df_item.columns = ['movieId','movieName']
df_item.head()
```

Following are the actions accomplished by each command in this cell:

a) `df_item = df_item.iloc[:,0:2]`: Removes all the columns in the DataFrame except for the first two columns that contain the movie ID and the movie title

b) `df_item.columns = ['movieId','movieName']`: Assigns names to the columns in the remaining columns in the DataFrame

c) `df_item.head()`: Displays the first few rows of the transformed DataFrame

This cell produces as output the first few rows of the transformed DataFrame, as shown in *Figure 5.14*:

	movieId	movieName
0	1	Toy Story (1995)
1	2	GoldenEye (1995)
2	3	Four Rooms (1995)
3	4	Get Shorty (1995)
4	5	Copycat (1995)

Figure 5.14 – The first few rows of the updated df_item DataFrame

13. Run the following cell to combine the `df_data` and `df_item` DataFrames into a single new DataFrame, `df`, that combines the columns from the original DataFrames:

```
df =\
pd.merge(df_data,df_item,on=['movieId'],how='left')
df.head()
```

Here are the arguments for `merge` to combine the DataFrames:

a) `df_data`: The DataFrame containing the user ID, movie ID, and rating.

b) `df_item`: The DataFrame containing the movie ID and movie title.

c) `on=['movieId']`: Specifies that the two DataFrames will be joined on the `movieID` column.

d) `how='left'`: Specifies that the rows from the `df_data` DataFrame will form the basis for the new DataFrame that is the result of the merge. That is, the new DataFrame will have the same number of rows as `df_data`, with each row including all the columns from `df_data` plus the `movieName` column from `df_item`.

This cell produces as output the first few rows of the new `df` DataFrame, as shown in *Figure 5.15*. Compared to `df_data`, you can see that `df` has one additional column, `movieName`:

	userId	movieId	rating	timestamp	movieName
0	196	242	3	881250949	Kolya (1996)
1	186	302	3	891717742	L.A. Confidential (1997)
2	22	377	1	878887116	Heavyweights (1994)
3	244	51	2	880606923	Legends of the Fall (1994)
4	166	346	1	886397596	Jackie Brown (1997)

Figure 5.15 – The first few rows of df

14. Run the following cell to see the dimensions of the `df` DataFrame:

```
df.shape
```

The output of this command shows the number of rows and columns in the `df` DataFrame. Note that `df` has the same number of rows as `df_data`:

```
(100000, 5)
```

Figure 5.16 – The shape of the df DataFrame

15. Run the following cell to see the number of unique values in each column of the `df` DataFrame:

```
df.nunique()
```

The output of this command lists all the columns of the DataFrame with the count of unique items in each column:

```
userId         943
movieId       1682
rating           5
timestamp    49282
movieName     1664
dtype: int64
```

Figure 5.17 – Count of unique values in each column of the df DataFrame

16. Run the following cell to see whether there are any missing values in any of the columns of `df`. If there are any missing values, we will need to deal with them prior to training the recommender system:

```
df.isnull().sum()
```

The output of this command lists the number of missing values for each column. Since there are no missing values, we can proceed with the steps to train the recommender system:

```
userId       0
movieId      0
rating       0
timestamp    0
movieName    0
dtype: int64
```

Figure 5.18 – Count of missing values in each column of the df DataFrame

17. Run the following cell to define a `CollabDataLoaders` object for the recommender system:

```
dls=/
CollabDataLoaders.from_df(df,item_name='movieName',bs=
64)
```

Here are the arguments for the definition of the `CollabDataLoaders` object:

a) `df`: The DataFrame used to create the `CollabDataLoaders` object

b) `item_name='movieName'`: Specifies the column that contains the name of the item that is the subject of the recommender system, in this case, `movieName`

c) `bs= 64`: Sets the batch size (the number of items on which the average loss is calculated) at 64

18. Run the following cell to see a batch from the `CollabDataLoaders` object that you defined in the previous step:

```
dls.show_batch()
```

The output displays the contents of the batch, as shown in *Figure 5.19*:

	userId	movieName	rating
0	459	Bogus (1996)	3
1	332	Time to Kill, A (1996)	5
2	458	Independence Day (ID4) (1996)	1
3	932	Bananas (1971)	4
4	130	Jerry Maguire (1996)	5
5	717	Executive Decision (1996)	4
6	776	Terminator, The (1984)	3
7	25	North by Northwest (1959)	4
8	116	Ulee's Gold (1997)	3
9	466	Jackie Chan's First Strike (1996)	3

Figure 5.19 – Output of show_batch

19. Run the following cell to define the model for the recommender system by defining a `collab_learner` object:

```
learn=collab_learner(dls,y_range= [ 1 , 5 ] )
```

Here are the arguments for the definition of the `collab_learner` object:

a) `dls`: The `CollabDataLoaders` object that you defined for the dataset.

b) `y_range= [ 1 , 5 ]`: Specifies the range of the values being predicted by the recommender system. In our case, this is the range of values for movie ratings in the `rating` column of the dataset.

20. Run the following cell to train the model for the recommender system:

```
learn.fit_one_cycle( 5 )
```

Here is the argument for the model definition:

a) 5: The number of epochs in the training run for the model

The output displays the training and validation loss for each epoch, as shown in *Figure 5.20*. You get an indication of the training improving as the validation loss decreases through the epochs:

epoch	train_loss	valid_loss	time
0	1.259702	1.244677	00:10
1	0.916582	0.937143	00:10
2	0.831102	0.894830	00:10
3	0.804616	0.880501	00:10
4	0.755278	0.878830	00:10

Figure 5.20 – Output of training the collaborative filtering model

21. Before we exercise the recommender system, let's get an idea of what ratings users give to a movie that has a reputation for not being very good. Run the following cell to see the subset of the `df` DataFrame for the movie `Showgirls`:

```
df_one_movie = df [df.movieName=='Showgirls (1995)']
df_one_movie.head()
```

The output shows the first few rows in the df DataFrame that contain ratings for *Showgirls*, as shown in *Figure 5.21*:

	userId	movieId	rating	timestamp	movieName
979	233	375	4	876374419	Showgirls (1995)
3969	201	375	3	884287140	Showgirls (1995)
6158	343	375	2	876406978	Showgirls (1995)
11568	291	375	1	874868791	Showgirls (1995)
31108	346	375	1	875266176	Showgirls (1995)

Figure 5.21 – Subset of the df DataFrame for a single movie

22. From the output of the previous cell, it seems like there is a range of ratings for this movie. Let's see what the average rating is for this movie for all users. Run the following cell to see the average rating for this movie:

```
df_one_movie['rating'].mean()
```

The output is shown in *Figure 5.22*. As we suspected, the average rating for this movie is indeed low:

1.9565217391304348

Figure 5.22 – Average rating for Showgirls

23. To exercise the recommender system, we want to get rating predictions for one user for a set of movies. Run the following cell to define a test DataFrame to exercise the trained recommender system:

```
scoring_columns = ['userId','movieId','movieName']
test_df = pd.DataFrame(columns=scoring_columns)
test_df.at[0,'userId'] = 607
test_df.at[0,'movieId'] = 242
test_df.at[0,'movieName'] = 'Kolya (1996)'
test_df.at[1,'userId'] = 607
test_df.at[1,'movieId'] = 302
test_df.at[1,'movieName'] = 'L.A. Confidential (1997)'
test_df.at[2,'userId'] = 607
test_df.at[2,'movieId'] = 375
test_df.at[2,'movieName'] = 'Showgirls (1995)'
test_df.head()
```

Here are the key elements of this cell:

a) `scoring_columns`: A list of the column names of the DataFrame

b) `test_df`: The DataFrame for containing the test entries with column names specified in the `scoring_columns` list

Note that the value for `userId` is the same for every row in this DataFrame. When we apply this DataFrame to our recommender system, we will be getting predicted ratings for one user for three movies.

The output of `test_df.head()` shows the contents of the completed DataFrame, as shown in *Figure 5.23*. For each row in this DataFrame, we will get the recommender system to predict the rating that `userID 607` would give for the movie in that row:

	userId	movieId	movieName
0	607	242	Kolya (1996)
1	607	302	L.A. Confidential (1997)
2	607	375	Showgirls (1995)

Figure 5.23 – Contents of test_df DataFrame

24. Run the following cell to get rating predictions from the recommender system for the entries in `test_df`:

```
dl = learn.dls.test_dl(test_df)
learn.get_preds(dl=dl)
```

The output of this cell shows the results of the recommender system, as shown in the following screenshot:

```
(tensor([4.2557, 4.3637, 2.2996]), None)
```

Figure 5.24 – Results of the recommender system

What do these results mean? The recommender system is predicting that `userID 607` will give the following ratings:

a) `4.2557` for `Kolya`

b) `4.3637` for `L.A. Confidential`

c) `2.2996` for `Showgirls`

This is not an exhaustive test of the recommender system, but the ratings seem plausible because a high rating is predicted for L.A. Confidential, a well-regarded movie, and a low rating is predicted for Showgirls, a movie that received very poor reviews.

Congratulations! You have trained a recommender system in fastai using a large curated dataset that includes a broad variety of rating combinations.

How it works...

In this section, you worked through a recipe to train a recommender system on a large dataset. It's worth summarizing how this recipe differed from the recipe in *Training a recommender system on a small curated dataset* section. Here are the key differences:

- **Structure of the datasets**: The small dataset consisted of a single file: ratings. csv. The large dataset is made up of over 20 files, although we only use two for the recipe in this section: u.data (which contains user IDs, movie IDs, and the user's ratings for the movies) and u.item (which contains additional information about movies, including their titles).

- **Size of the datasets**: The small dataset has a little over 6,000 records. The large dataset has 100 k records.

- **Test DataFrame**: For the recommender system trained on a small dataset, the test DataFrame contained the user IDs and movie IDs for which we wanted to get predicted ratings. For the recommender system trained on a large dataset, the test DataFrame contained user IDs, movie IDs, and movie titles for which we wanted to get predicted ratings.

The facilities of fastai make it easy to create recommender systems on both the small and large curated datasets. In the next section, we'll explore how to use fastai to create a recommender system on a standalone dataset.

Training a recommender system on a standalone dataset

In the *Training a recommender system on a small curated dataset* and *Training a recommender system on a large curated dataset* recipes, we used two curated datasets adapted from the MovieLens dataset that contains user IDs, movie IDs, and user ratings for these movies. Using these two curated datasets, we created recommender systems that predicted what rating a user would give to a particular movie.

In this section, we will explore a dataset that is not a part of fastai's set of curated datasets. The dataset we will use in this section is the Amazon product dataset (`https://www.kaggle.com/saurav9786/amazon-product-reviews`). This dataset contains user ratings for a large range of products available on Amazon.

In this section, we will be working with a subset of the dataset, the subset related to electronic goods. In the recipe in this section, we will ingest this standalone dataset and then use it to train a recommender system that can predict a user's rating for an item.

Getting ready

Confirm that you can open the `training_recommender_systems_on_standalone_dataset.ipynb` notebook in the `ch5` directory of your repository.

Ensure that you have uploaded the file for the Amazon product dataset to your Gradient environment by following these steps:

1. Download the `archive.zip` file from `https://www.kaggle.com/saurav9786/amazon-product-reviews`.

2. Unzip the downloaded `archive.zip` file to extract `ratings_Electronics (1).csv`. Rename this file to `ratings_Electronics.csv`.

3. From the terminal in your Gradient environment, make `/storage/archive` your current directory:

   ```
   cd /storage/archive
   ```

4. Create the `/storage/archive/amazon_reviews` directory:

   ```
   mkdir amazon_reviews
   ```

5. Make `/storage/archive/amazon_reviews` your current directory:

   ```
   cd /storage/archive/amazon_reviews
   ```

6. Upload the files you extracted in *step 2* (`ratings_Electronics.csv`) to `/storage/archive/amazon_reviews`. You can use the upload button in JupyterLab in Gradient to do the upload by following these steps:

   a) From the terminal in your Gradient environment, make `/notebooks` your current directory:

   ```
   cd /notebooks
   ```

b) If you have not already created a `notebooks/temp` directory, make a new `/notebooks/temp` directory:

```
mkdir temp
```

c) In the JupyterLab file browser, make `temp` your current folder, select the upload button (see *Figure 5.25*), and then select the `ratings_Electronics.csv` file from your local system folder where you extracted it in *step 2*:

Figure 5.25 – The upload button in JupyterLab

d) From the terminal in your Gradient environment, copy the `ratings_Electronics.csv` file into the `/storage/archive/amazon_reviews` directory:

```
cp /notebooks/temp/ratings_Electronics.csv /storage/
archive/amazon_reviews/ratings_Electronics.csv
```

I gratefully acknowledge the opportunity to use the Amazon product dataset to illustrate the recommender system capabilities of fastai on non-curated datasets.

> **Dataset citation**
>
> J. McAuley, C. Targett, J. Shi, A. van den Hengel (2015). *Image-based recommendations on styles and substitutes* (http://cseweb.ucsd.edu/~jmcauley/pdfs/sigir15.pdf). Special Interest Group on Information Retrieval (SIGIR 2015)

Now that you have copied the dataset file into your Gradient environment, you are ready to go through the recipe.

How to do it...

In this section, you will be running through the `training_recommender_systems_on_standalone_dataset.ipynb` notebook. Once you have the notebook open in your fastai environment, complete the following steps:

1. Run the cells in the notebook up to the `Ingest the dataset` cell to import the required libraries and set up your notebook.

2. Run the following cell to define the path object for this dataset:

   ```
   path = URLs.path('amazon_reviews')
   ```

   The argument `'amazon_reviews'` indicates that this path object is being defined to point to the `/storage/archive/amazon_reviews` directory that you created in the *Getting ready* section.

3. Run the following cell to examine the directory structure of the dataset:

   ```
   path.ls()
   ```

   The output shows the directory structure of the dataset, as shown in *Figure 5.26*:

   ```
   (#1) [Path('/storage/archive/amazon_reviews/ratings_Electronics.csv')]
   ```

   Figure 5.26 – The output of path.ls() for the Amazon product dataset

4. Run the following cell to define the `df` DataFrame to contain the contents of the `ratings_Electronics.csv` file:

   ```
   df = pd.read_csv(path/'ratings_Electronics.csv',header = None)
   ```

   Here are the arguments for the definition of the `df` DataFrame:

   a) `path/'ratings_Electronics.csv'`: Specifies the source for this DataFrame, the `ratings_Electronics.csv` file

   b) `header = None`: Specifies that this file does not include column names in the first row

5. Run the following cell to define names for the columns of the `df` DataFrame. These column names come from the description of the dataset at `https://www.kaggle.com/saurav9786/amazon-product-reviews`:

   ```
   df.columns = ['userID','productID','rating','timestamp']
   ```

6. Run the following cell to examine the df DataFrame:

```
df.head()
```

The output, as shown in *Figure 5.26*, lists records from the dataset. Each record represents a user's rating for a product. The userID column has the ID for the user. The productID column has the ID for the product. The rating column is the rating given by the user for the product:

	userID	productID	rating	timestamp
0	AKM1MP6P0OYPR	0132793040	5.0	1365811200
1	A2CX7LUOHB2NDG	0321732944	5.0	1341100800
2	A2NWSAGRHCP8N5	0439886341	1.0	1367193600
3	A2WNBOD3WNDNKT	0439886341	3.0	1374451200
4	A1GIOU4ZRJA8WN	0439886341	1.0	1334707200

Figure 5.27 – Output of df.head()

7. Run the following cell to get the dimensions of the df DataFrame:

```
df.shape
```

The output, as shown in *Figure 5.27*, shows that the DataFrame has over 7 million rows:

$$(7824482, 4)$$

Figure 5.28 – Shape of DataFrame df

8. Run the following cell to define the CollabDataLoaders object, dls:

```
dls=CollabDataLoaders.from_df(df,bs= 64)
```

Here are the arguments to the definition of the CollabDataLoaders object, dls:

a) df: Specifies that the DataFrame you created for the dataset, df, is used to create the CollabDataLoaders object

b) bs = 64: Specifies that the batch size is 64

9. Run the following cell to see a batch from the CollabDataLoaders object that you defined in the previous step:

```
dls.show_batch()
```

The output displays the contents of the batch, as shown in *Figure 5.28*:

	userID	productID	rating
0	APSVFXSVU0P6C	B008ABOJKS	4.0
1	A2C2TQICKW8W8	B00BQ4SBSM	2.0
2	A1I2HYPP41PIAF	B000BKJZ9Q	2.0
3	A6XLG77BC9R8R	B003A4H4VQ	5.0
4	A2NYK9KWFMJV4Y	B002JDVBYU	5.0
5	A1H9OR8UASFIR6	B000BMAQAQ	4.0
6	A1E7USO8M79M7A	B0018JV6X2	1.0
7	A1N6RWK9XBXG3T	B007B31IYQ	5.0
8	A5NBOXDPQ75RJ	B006202R44	5.0
9	A29ZTEO6EKSRDV	B004S4XNKI	3.0

Figure 5.29 – Output of show_batch()

10. Run the following cell to define the `collab_learner` object for the recommender system model:

```
learn=collab_learner(dls,y_range= [ 0 , 5.0 ] )
```

Here are the arguments for the definition of the `collab_learner` object:

a) `dls`: The `CollabDataLoaders` object that you defined in a previous step

b) `y_range`: The range of values in the `rating` column

11. Run the following cell to train the recommender system model:

```
learn.fit_one_cycle( 1 )
```

Here is the argument for the model definition:

a) `1`: The number of epochs in the training run for the model.

> **Note**
>
> The dataset you are using in this recipe is big. Recall that when you ran the cell to get the dimensions of the `df` DataFrame, you found that this DataFrame has over 7 million rows. What this means is that training the model will take some time. For example, it took me over 3 hours to train a recommender system model on this dataset in a Gradient environment.

The output shows the training and validation loss, as shown in the following screenshot:

epoch	train_loss	valid_loss	time
0	2.829242	2.826828	3:36:08

Figure 5.30 – Output of training the collaborative filtering model

12. Run the following cell to define the `test_df` DataFrame, which includes a couple of test entries that you can use to exercise the trained recommender system:

```
scoring_columns = ['userID','productID']
test_df = pd.DataFrame(columns=scoring_columns)
test_df.at[0,'userID'] = 'A2NYK9KWFMJV4Y'
test_df.at[0,'productID'] = 'B008ABOJKS'
test_df.at[1,'userID'] = 'A29ZTEO6EKSRDV'
test_df.at[1,'productID'] = 'B006202R44'
test_df.head()
```

Here are the key elements of this cell:

a) `scoring_columns`: A list of the column names of the DataFrame. Note that these column names are the same column names that you saw in the output of `show_batch()` in *Figure 5.29*.

b) `test_df`: The DataFrame for containing the test entries with column names specified in the `scoring_columns` list.

The output of `test_df.head()` shows the contents of the completed `test_df` DataFrame, as shown in the following screenshot:

	userID	productID
0	A2NYK9KWFMJV4Y	B008ABOJKS
1	A29ZTEO6EKSRDV	B006202R44

Figure 5.31 – Contents of the test_df DataFrame

13. Now that you have defined the `test_df` DataFrame, you can use it to exercise the trained recommender system. Run the following cell to get rating predictions from the recommender system for the entries in `test_df`:

```
dl = learn.dls.test_dl(test_df)
learn.get_preds(dl=dl)
```

The output of this cell shows the results of the recommender system, as shown in the following screenshot:

```
(tensor([4.4364, 2.5531]), None)
```

Figure 5.32 – Results of the recommender system

What do these results mean?

a) The first entry in the output, 4.4364, is the recommender system's prediction for the first entry in `test_df`. The recommender system is predicting that the user with a user ID of A2NYK9KWFMJV4Y will give a rating of 4.4364 for the product with a product ID of B008ABOJKS.

b) The second entry in the output, 2.5531, is the recommender system's prediction for the second entry in `test_df`. The recommender system is predicting that the user with a user ID of A29ZTEO6EKSRDV will give a rating of 2.5531 for the product with a product ID of B006202R44.

Congratulations! You have trained a recommender system in fastai using a standalone dataset.

How it works...

In this section, you worked through a recipe for training a recommender system on a standalone dataset. Like the standalone datasets that you encountered in *Chapter 3, Training Models with Tabular Data*, and *Chapter 4, Training Models with Text Data*, the standalone dataset that you used in this section required you to take some additional steps compared to the fastai curated datasets that you used in the *Training a recommender system on a small curated dataset* and *Training a recommender system on a large curated dataset* recipes.

Here is a summary of the additional steps required for a standalone dataset:

1. **Find a dataset**. This may sound simplistic, but once you move beyond the world of the fastai curated datasets and try to learn more by using standalone datasets, it can be a challenge to find the right dataset. The Kaggle site, https://www.kaggle.com/, is a great place to start. You want a dataset that's big enough to give a fighting chance when it comes to training a deep learning model.

My experience has been that if I'm not using transfer learning (that is, starting off with a pre-existing model that has been trained on a general dataset that applies to the use case I want to train my deep learning model on), then I need to have a dataset that has at least several tens of thousands of items in it. .

You want a dataset that's not too big. As you saw in the recipe in this section and in the language models you trained in *Chapter 4, Training Models with Text Data*, training a model with a big dataset can take hours. If you're using a Gradient instance with a cost (as opposed to Colab or a free Gradient instance), you can end up spending several dollars for a single training run. Costs can add up if you need to do multiple training runs.

2. **Bring the dataset into your fastai environment**. As described in the *Getting ready* section of this recipe, after you have downloaded the dataset to your local system, you have to create a directory in your fastai environment to hold the dataset files and then upload the files to that directory.

One of the advantages of fastai is the large number of curated datasets. However, for dataset types such as recommender systems, where fastai only offers one or two curated datasets, it is advantageous for you to feel comfortable with finding standalone datasets with different characteristics, so you can perform a wider variety of experiments. By following the recommendations in this section, you will be able to expand your world of datasets beyond the fastai curated datasets when you're ready to learn more.

Test your knowledge

Now that you have worked through some extended examples of training fastai recommender systems, you can try some variations to exercise what you've learned.

Getting ready

Ensure that you have followed the *Training a recommender system on a standalone dataset* recipe. In this section, you will be adapting the notebook worked through in that recipe to create a recommender system for a new standalone dataset.

How to do it...

You can follow the steps in this section to try some variations on the recommender system that you trained with the Amazon product dataset in the *Training a recommender system on a standalone dataset* recipe:

1. Make a copy of the `training_recommender_systems_on_standalone_dataset.ipynb` notebook that you worked through in the *Training a recommender system on a standalone dataset* recipe. Give your new copy of the notebook the following name: `training_recommender_systems_on_new_standalone_dataset.ipynb`.

2. Review the description of the whole Amazon product dataset at `http://jmcauley.ucsd.edu/data/amazon/`. From the list of **small subsets for experimentation**, select a category other than **Electronics**. You have already trained a recommender system with the **Electronics** dataset in the *Training a recommender system on a standalone dataset* recipe, so you will want to pick another category.

   For example, you could pick **Office Products** or **Automotive**. I suggest avoiding the **Books** category because its file is three times larger than the **Electronics** dataset, so you could be facing a whole day training a recommender system on the *Books* dataset.

3. Adapt the steps in the *Getting ready* section of the *Training a recommender system on a standalone dataset* recipe to get the *ratings only* dataset for the category you chose in *Step 2* in your fastai environment.

4. Update your `training_recommender_systems_on_new_standalone_dataset.ipynb` notebook to ingest the dataset you brought into your fastai environment in *Step 3*.

5. Update your notebook to load the dataset into a DataFrame and then use the techniques you have seen in this chapter (including `head()` and `shape`) to examine the structure of the dataset.

6. Using what you learned in *Step 5*, update the rest of your `training_recommender_systems_on_new_standalone_dataset.ipynb` notebook to create, train, and test a recommender system for the dataset you selected in *Step 2*.

Congratulations! You have completed a review of training recommender systems using fastai.

6
Training Models with Visual Data

Deep learning has been successfully applied to many different types of data, including tabular data, text data, and recommender system data. You saw fastai's approach to these types of data in *Chapter 3*, *Training Models with Tabular Data*, *Chapter 4*, *Training Models with Text Data*, and *Chapter 5*, *Training Recommender Systems*. These types of data are all part of the story of deep learning, but **visual data** or **image data** is the type of data that is traditionally associated most closely with deep learning.

Visual data is also the type of data that is most thoroughly supported by the fastai framework. The fastai high-level API is mostly developed for visual data, and 70% of the curated fastai datasets are visual datasets. In this chapter, we will explore some of the features that fastai provides for exploring visual datasets and building high-performance deep learning models with image datasets.

In this chapter, you will learn how to use the rich set of facilities available in fastai for preparing image datasets and using them to train deep learning models. In particular, you will learn how to create fastai deep learning models that classify images, that is, determine what objects are in images, and also how to use fastai to identify multiple objects in the same image.

Here are the recipes that will be covered in this chapter:

- Training a classification model with a simple curated vision dataset
- Exploring a curated image location dataset
- Training a classification model with a standalone vision dataset
- Training a multi-image classification model with a curated vision dataset
- Test your knowledge

Technical requirements

Ensure that you have completed the setup sections from *Chapter 1*, *Getting Started with fastai*, and have a working Gradient instance or Colab set up. The recipes described in this chapter assume that you are using Gradient. Ensure that you have cloned the repo for this book from `https://github.com/PacktPublishing/Deep-Learning-with-fastai-Cookbook` and have access to the `ch6` folder. This folder contains the code samples described in this chapter.

Training a classification model with a simple curated vision dataset

You may recall the first fastai model that you trained back in *Chapter 1*, *Getting Started with fastai*. That model was trained on the MNIST dataset of hand-written digits. Given an image of a hand-written digit, that model was able to classify the image, that is, determine which of the digits from 0 to 9 were shown in the image.

In this recipe, you are going to apply the same approach you saw in the MNIST model to another fastai curated dataset: the CIFAR dataset. This dataset, which is a subset of a larger curated CIFAR_100 dataset, is made up of 6,000 images organized into 10 categories. The model that you train in this section will be able to determine the category that an image from this dataset belongs to.

Getting ready

Confirm that you can open the `training_with_curated_image_datasets.ipynb` notebook in the `ch6` directory of your repo.

> **Note**
>
> The images in the CIFAR dataset are quite small. In this section, we have rendered them in a larger size to make them easier to recognize in the context of the book, but the outcome is that they can look a bit blurry.

The CIFAR dataset featured in this section is introduced in the paper *Learning Multiple Layers of Features from Tiny Image*, Krizhevsky, 2009. I am grateful for the opportunity to include this dataset in this book.

> **Dataset citation**
>
> Alex Krizhevsky. (2009). *Learning Multiple Layers of Features from Tiny Image*. University of Toronto: https://www.cs.toronto. edu/~kriz/learning-features-2009-TR.pdf.

How to do it...

In this section, you will be running through the training_with_curated_ image_datasets.ipynb notebook. Once you have the notebook open in your fastai environment, complete the following steps:

1. Run the cells in the notebook up to the Ingest the dataset cell to import the required libraries and set up your notebook.

2. Run the following cell to define the path object for this dataset:

```
path = untar_data(URLs.CIFAR)
```

3. Run the following cell to examine the directory structure of the dataset:

```
path.ls()
```

The output shows the directory structure of the dataset, as shown in the following screenshot:

```
(#3) [Path('/storage/data/cifar10/test'),Path('/storage/data/cifar10/train'),Path('/storage/data/cifar1
0/labels.txt')]
```

Figure 6.1 – Output of path.ls()

4. Run the following cell to define an ImageDataLoaders object for this dataset:

```
dls = ImageDataLoaders.from_folder(path, train='train',
      valid='test')
```

Here are the arguments for the definition of the `ImageDataLoaders` object:

a) `path`: Specifies that the `ImageDataLoaders` object is defined using the `path` object you created in the previous cell

b) `train='train'`: Specifies that the training data is in the `/storage/data/cifar10/train` directory

c) `valid='test'`: Specifies that the validation data is in the `/storage/data/cifar10/test` directory

5. Run the following cell to display a batch from the dataset:

```
dls.train.show_batch(max_n=4, nrows=1)
```

The output of this cell is a set of 4 items from a batch, showing images along with their corresponding categories, as shown in the following figures:

Figure 6.2 – Output of show_batch()

6. Run the following cell to examine the contents of the `train` subdirectory:

```
(path/'train').ls()
```

The output shows the structure of the `train` subdirectory, as shown in the following screenshot:

```
(#10) [Path('/storage/data/cifar10/train/dog'),Path('/storage/data/cifar10/train/automobile'),Path('/st
orage/data/cifar10/train/frog'),Path('/storage/data/cifar10/train/airplane'),Path('/storage/data/cifar1
0/train/deer'),Path('/storage/data/cifar10/train/horse'),Path('/storage/data/cifar10/train/cat'),Path
('/storage/data/cifar10/train/truck'),Path('/storage/data/cifar10/train/bird'),Path('/storage/data/cifa
r10/train/ship')]
```

Figure 6.3 – Contents of the train subdirectory

7. Run the following cell to examine the contents of the `train/dog` subdirectory:

```
(path/'train/dog').ls()
```

The output shows the structure of the `train/dog` subdirectory:

```
(#5000) [Path('/storage/data/cifar10/train/dog/15233_dog.png'),Path('/storage/data/cifar10/train/dog/15
19_dog.png'),Path('/storage/data/cifar10/train/dog/13990_dog.png'),Path('/storage/data/cifar10/train/do
g/16280_dog.png'),Path('/storage/data/cifar10/train/dog/42635_dog.png'),Path('/storage/data/cifar10/tra
in/dog/37152_dog.png'),Path('/storage/data/cifar10/train/dog/16540_dog.png'),Path('/storage/data/cifar1
0/train/dog/2182_dog.png'),Path('/storage/data/cifar10/train/dog/48048_dog.png'),Path('/storage/data/ci
far10/train/dog/17186_dog.png')...]
```

Figure 6.4 – Contents of the train/dog subdirectory

8. If you want to take a different perspective as regards the directory structure of this dataset, you can use the `tree` command. To do this, from the Gradient terminal, enter the following commands:

```
cd /storage/data/cifar10
tree -d
```

The output of the command shows the structure of the dataset. You can see that each of the `test` and `train` directories have subdirectories for each of the 10 categories of the dataset:

```
├── test
│   ├── airplane
│   ├── automobile
│   ├── bird
│   ├── cat
│   ├── deer
│   ├── dog
│   ├── frog
│   ├── horse
│   ├── ship
│   └── truck
└── train
    ├── airplane
    ├── automobile
    ├── bird
    ├── cat
    ├── deer
    ├── dog
    ├── frog
    ├── horse
    ├── ship
    └── truck
```

9. Run the following cell to view a single item in the dataset:

```
img_files = get_image_files(path)
img = PILImage.create(img_files[100])
img
```

Here are the key elements of this cell:

a) `img_files = get_image_files(path)`: Specifies that `path` is recursively examined and returns all the image files in the path. If you want more details about `get_image_files`, you can check out the documentation at the following link: `https://docs.fast.ai/data.transforms.html#get_image_files`.

b) `img = PILImage.create(img_files[100])`: Creates the image object, `img`, from a specific file returned by the previous statement.

The output of this cell is one of the dataset files rendered as an image in the notebook, as shown here:

Figure 6.5 – An image from the dataset

10. Run the following cell to display another image from the dataset:

```
img = PILImage.create(img_files[3000])
img
```

The output is another of the dataset files rendered as an image in the notebook as follows:

Figure 6.6 – Another image from the dataset

11. Run the following cell to define the model as a `cnn_learner` object:

```
learn = cnn_learner(dls, resnet18,
                    loss_
func=LabelSmoothingCrossEntropy(),
                    metrics=accuracy))
```

Here are the arguments for the definition of the `cnn_learner` object:

a) `dls`: Specifies that the model uses the `ImageDataLoaders` object, `dls`, that you defined in *Step 4*

b) `resnet18`: Specifies the pre-trained model to use as a starting point for this model

c) `loss_func=LabelSmoothingCrossEntropy()`: Specifies the loss function to use in the training process

d) `metrics=accuracy`: Specifies that `accuracy` is the performance metric that will be optimized in the training process

12. Run the following cell to train the model:

```
learn.fine_tune(5)
```

The argument indicates that the training run will be for 5 epochs.

The output displays the training loss, validation loss, and accuracy for each epoch, as shown in the following screenshot:

epoch	train_loss	valid_loss	accuracy	time
0	1.834696	1.698438	0.467800	00:56

epoch	train_loss	valid_loss	accuracy	time
0	1.312279	1.225815	0.677300	01:05
1	1.111695	1.048064	0.760600	01:06
2	0.947037	0.991742	0.785300	01:05
3	0.810161	0.987946	0.791500	01:05
4	0.745295	0.995719	0.793400	01:05

Figure 6.7 – Output of training the model

13. Let's try out the trained model on some examples from the test dataset. First, run the following cell to define an object for one of the images in the test dataset:

```
img_test_files = get_image_files(path/"test")
img2 = PILImage.create(img_test_files[700])
img2
```

Here are the key elements of this cell:

a) `img_files = get_image_files(path/"test"))`: Returns all the image files under the `test` directory

b) `img = PILImage.create(img_files[700])`: Creates the image object, `img2`, from a specific file returned by the previous statement

The output of this cell is an image of a dog:

Figure 6.8 – Dog image from the test dataset

14. Run the following cell to define an object for another one of the images in the test dataset:

```
img3 = PILImage.create(img_test_files[8000])
img3
```

The output of this cell is an image of a bird:

Figure 6.9 – Bird image from the test dataset

15. Now that we have defined objects for a couple of images from the test dataset, let's exercise the trained image classification model on them. First, run the following cell to apply the model to the dog image:

```
learn.predict(img2)
```

The output of this cell is the model's prediction, as shown in the following screenshot. Note that the model correctly predicts the category of the image:

```
('dog',
 TensorImage(5),
 TensorImage([0.0167, 0.0154, 0.1496, 0.0879, 0.0338, 0.4890, 0.1772, 0.0096, 0.0082, 0.0125]))
```

Figure 6.10 – Image classification model's prediction on the dog image

16. Let's now see how the model does with the image of a bird. Run the following cell to apply the model to the bird image:

```
learn.predict(img3)
```

The output of this cell is the model's prediction, as shown in the following screenshot. Note that the model correctly predicts that the image is a bird:

```
('bird',
 TensorImage(2),
 TensorImage([0.0153, 0.0173, 0.7091, 0.0153, 0.1924, 0.0120, 0.0116, 0.0098, 0.0067, 0.0105]))
```

Figure 6.11 – Image classification model's prediction on the bird image

17. Now that we have successfully exercised the model, let's save it. Run the following cell to save the model:

```
learn.path = Path('/notebooks/temp')
learn.export('cifar_apr20_2021.pkl')
```

Here is what the statements in this cell do:

a) `learn.path = Path('/notebooks/temp')`: Sets the path of the `learn` object to a directory that can be written to. Remember that in Gradient, by default, the path for the `learn` object is read-only, so you need to adjust the path to a writeable directory before you can save the model.

b) `learn.export('cifar_apr20_2021.pkl')`: Specifies that the name of the model to be saved is `cifar_apr20_2021.pkl`.

After you have run this cell in Gradient, your model is saved in `/notebooks/temp/model/cifar_apr20_2021.pkl`.

Congratulations! You have trained and exercised a fastai image classification model using a curated image dataset.

How it works...

Some important things are going on in the recipe in this section that are worth reviewing. In this section, we'll go over some of the details that might not have been evident in the main part of the recipe.

Labels are encoded in directory names and filenames

First, consider the names of the files that make up the dataset. The files in the `train/dog` subdirectory are shown here:

```
(#5000) [Path('/storage/data/cifar10/train/dog/15233_dog.png'),Path('/storage/data/cifar10/train/dog/15
19_dog.png'),Path('/storage/data/cifar10/train/dog/13990_dog.png'),Path('/storage/data/cifar10/train/do
g/16280_dog.png'),Path('/storage/data/cifar10/train/dog/42635_dog.png'),Path('/storage/data/cifar10/tra
in/dog/37152_dog.png'),Path('/storage/data/cifar10/train/dog/16540_dog.png'),Path('/storage/data/cifar1
0/train/dog/2182_dog.png'),Path('/storage/data/cifar10/train/dog/48048_dog.png'),Path('/storage/data/ci
far10/train/dog/17186_dog.png')...]
```

Figure 6.12 – Files in the train/dog subdirectory

The files in the `train/cat` subdirectory are shown here:

```
(#5000) [Path('/storage/data/cifar10/train/cat/42930_cat.png'),Path('/storage/data/cifar10/train/cat/21
351_cat.png'),Path('/storage/data/cifar10/train/cat/4151_cat.png'),Path('/storage/data/cifar10/train/ca
t/34987_cat.png'),Path('/storage/data/cifar10/train/cat/2197_cat.png'),Path('/storage/data/cifar10/trai
n/cat/15181_cat.png'),Path('/storage/data/cifar10/train/cat/26462_cat.png'),Path('/storage/data/cifar10
/train/cat/38932_cat.png'),Path('/storage/data/cifar10/train/cat/19061_cat.png'),Path('/storage/data/ci
far10/train/cat/23239_cat.png')...]
```

Figure 6.13 – Files in the train/cat subdirectory

Let's take a look at a sample file from each of these subdirectories. To display an image file from the `train/dog` subdirectory, run the following cell:

```
dog_files = get_image_files(path/"train/dog")
dog_img = PILImage.create(dog_files[30])
dog_img
```

The output of this cell is indeed an image of a dog, as shown here:

Figure 6.14 – A dog image from the training dataset

To display an image file from the `train/cat` subdirectory, run the following cell:

```
cat_files = get_image_files(path/"train/cat")
cat_img = PILImage.create(cat_files[30])
cat_img
```

The output of this cell is, as expected, an image of a cat, as shown here:

Figure 6.15 – A cat image from the training dataset

For this image classification problem, the model is trying to predict the label for an image, that is, the category, such as dog, deer, or bird, of the object that is shown in the image. For this dataset, the label is encoded in the following two ways:

- **The name of the directory containing the image file**. As you can see in *Figure 6.14* and *Figure 6.15*, the label for an image is encoded in the name of the subdirectory where the image is located.

- **The name of the file**. As you can see in *Figure 6.12* and *Figure 6.13*, the dog images all have filenames of the form xxxx_dog.png, and the cat images have filenames of the form xxxx_cat.png.

You used transfer learning to train the image classification model

In *Chapter 4, Training Models with Text Data*, we used transfer learning to adapt an existing trained model to work with a dataset for a particular use case. You may not have noticed it, but we did the same thing in this recipe. There's a clue in the cell where you defined the cnn_learner object as follows:

```
learn = cnn_learner(dls, resnet18,
                    loss_func=LabelSmoothingCrossEntropy(),
                    metrics=accuracy))
```

In the definition of cnn_learner, the resnet18 argument is the pre-trained model that is used as a basis for the image classification model. The first time you run this cell, you will see a message like the one shown in *Figure 6.16*, indicating that the model is being set up in your environment:

```
Downloading: "https://download.pytorch.org/models/resnet18-5c106cde.pth" to /root/.cache/torch/hub/checkpoints/resnet
18-5c106cde.pth
```

Figure 6.16 – Message you get the first time you run the cnn_learner definition

The second clue is in the cell where you train the model, as follows:

```
learn.fine_tune(5)
```

For most of the models you have seen so far in this book, you trained the model using learn.fit_one_cycle(). Here, because we want to update an existing model for our particular use case, we use learn.fine_tune() instead. This is exactly what we did in *Chapter 4*, *Training Models with Text Data*, for the language models we trained using transfer learning on top of the pre-trained AWD_LSTM model.

Why did we use transfer learning for this use case instead of training the model from scratch? The simple answer is that we get decent performance from the model faster by using transfer learning. You can try it out yourself by making two changes to the training_with_curated_image_datasets.ipynb notebook and rerunning it. Here are the steps to follow to do this:

1. Update the definition of the cnn_learner object to include the pretrained=False argument, as shown here:

    ```
    learn = cnn_learner(dls, resnet18, pretrained=False,
                        loss_
    func=LabelSmoothingCrossEntropy(), metrics=accuracy)
    ```

    This change means that the model will be trained from scratch with the CIFAR dataset rather than by taking advantage of the pre-trained model.

2. Change the training statement to the following:

    ```
    learn.fit_one_cycle(5)
    ```

When you run the notebook after making these changes, this new model will not do a very good job of classifying images. Transfer learning works, and as Howard and Gugger explain in their book, it doesn't get nearly enough attention in standard introductions to deep learning. It's lucky for us that fastai is designed to make it easy to exploit the power of transfer learning, as shown by the image classification model that you trained in the *How to do it…* section of this recipe.

A closer look at the model

Before moving on to the next recipe, it's worth taking a closer look at the model from this recipe. A deeply detailed description of the model is beyond the scope of this book, so we will just focus on some highlights here.

As shown in the recipe, the model is defined as a `cnn_learner` object (documentation here: `https://docs.fast.ai/vision.learner.html#cnn_learner`) that uses the pre-trained `resnet18` model (documentation here: `https://pytorch.org/vision/stable/models.html`). The output of `learn.summary()` for this model shows that the model includes a series of convolutional layers and associated layers. For a description of convolutional neural networks (CNNs), see: `https://machinelearningmastery.com/how-to-develop-a-convolutional-neural-network-from-scratch-for-mnist-handwritten-digit-classification/`.

Exploring a curated image location dataset

Back in *Chapter 2, Exploring and Cleaning Up Data with fastai*, we went through the process of ingesting and exploring a variety of datasets using fastai.

In this section, we are going to explore a special curated image dataset called `COCO_TINY`. This is an **image location** dataset. Unlike the `CIFAR` dataset that we used in the *Training a classification model with a simple curated vision dataset* recipe, which had a single labeled object in each image, the images in image location datasets are labeled with bounding boxes (which indicate where in the image a particular object occurs) as well as the name of the object. Furthermore, images in the `COCO_TINY` dataset can contain multiple labeled objects, as shown here:

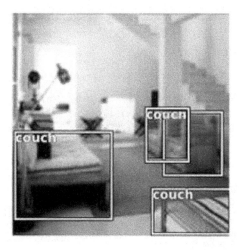

Figure 6.17 – Labeled image from an image location dataset

In the recipe in this section, we'll ingest the dataset and apply its annotation information to create a `dataloaders` object for the dataset. The annotation information for each image can be quite complex. Each image can depict multiple labeled objects. For each labeled object in an image, the annotation information includes the filename of the image, the *x* and *y* coordinates of the bounding box that surrounds the object, and the label for the category of the object. For example, the image shown in *Figure 6.17* contains several **couch** objects and a **chair** object.

Note that the images in the dataset don't actually show the annotation. A typical image from the dataset looks as follows:

Figure 6.18 – Raw image from the COCO_TINY dataset

There are no bounding boxes or labels in the image itself. That annotation information is contained in a separate file. The recipe in this section will show you how to combine the image files with the annotation information.

At the end of this recipe, when we use `show_batch()` to display samples from the batch that incorporates annotation, the bounding boxes and labels are shown in the images. For example, *Figure 6.19* shows how the image from *Figure 6.18* looks when it is displayed by `show_batch()` – now you can see the bounding box and label for the **vase** object in the image:

Figure 6.19 – Annotated version of the image in Figure 6.18

In the recipe in this section, you will combine the image files from the `COCO_TINY` dataset, such as the image shown in *Figure 6.18*, with the information from the annotation file to get a `dataloaders` object that includes the images along with bounding boxes and labels for all the labeled objects in each image, as shown in *Figure 6.19*.

Getting ready

Confirm that you can open the `exploring_image_location_datasets.ipynb` notebook in the `ch6` directory of your repository.

The approach to examining the dataset shown in this recipe includes approaches inspired by this Kaggle kernel, `https://www.kaggle.com/jachen36/coco-tiny-test-prediction`, which demonstrates how to use fastai to explore an image location dataset.

The COCO_TINY dataset featured in this section is introduced in the paper *Microsoft COCO: Common Objects in Context*, Lin et al., 2014. I am grateful for the opportunity to include an example using this dataset in this book.

> **Dataset citation**
>
> Tsung-Yi Lin, Michael Maire, Serge Belongie, Lubomir Bourdev, Ross Girshick, James Hays, Pietro Perona, Deva Ramanan, C. Lawrence Zitnick, and Piotr Dollár. (2014). *Microsoft COCO: Common Objects in Context*: https:// arxiv.org/abs/1405.0312.

How to do it...

In this section, you will be running through the exploring_image_location_ datasets.ipynb notebook. Once you have the notebook open in your fastai environment, complete the following steps:

1. Run the cells in the notebook up to the Ingest the dataset cell to import the required libraries and set up your notebook.

2. Run the following cell to define the path object for this dataset:

    ```
    path = untar_data(URLs.COCO_TINY)
    ```

3. Run the following cell to examine the directory structure of the dataset:

    ```
    path.ls()
    ```

    The output shows the directory structure of the dataset, as shown in *Figure 6.20*:

    ```
    (#2) [Path('/storage/data/coco_tiny/train'),Path('/storage/data/coco_tiny/train.json')]
    ```

    Figure 6.20 – Output of path.ls()

    The dataset consists of a set of image files in the train subdirectory and the train.json annotation file. In the next few steps of this recipe, we will take a closer look at what's in the train.json file.

4. Run the following cell to bring the train.json file into a series of Python dictionaries:

    ```
    with open(path/'train.json') as json_file:
        data = json.load(json_file)
        # each nested structure is a list of dictionaries
        categories = data['categories']
    ```

```
images = data['images']
annotations = data['annotations']
```

Here are the key parts of the code used in this cell:

a) `data = json.load(json_file)`: Loads the contents of the whole `train.json` file into the `data` dictionary.

b) `categories = data['categories']`: Creates a separate list of dictionaries just for the category definitions. This dictionary defines the objects in the images.

c) `images = data['images']`: Creates a separate list of dictionaries just for image files.

d) `annotations = data['annotations']`: Creates a separate list of dictionaries just for the bounding boxes.

5.   Run the following cell to see the structure of each of the dictionaries you created in the previous cell:

```
print("categories ", categories)
print()
print("subset of images",list(images)[:5])
print()
print("subset of annotations",list(annotations)[:5])
```

The output of this cell shows samples of the contents of each of the three dictionaries, as shown in the following screenshot:

```
categories  [{'id': 62, 'name': 'chair'}, {'id': 63, 'name': 'couch'}, {'id': 72, 'name': 'tv'}, {'id': 75, 'name': 'remote'}, {'id': 84, 'name': 'book'}, {'id': 86, 'name': 'vase'}]

subset of images [{'id': 542959, 'file_name': '000000542959.jpg'}, {'id': 129739, 'file_name': '0000001 29739.jpg'}, {'id': 153607, 'file_name': '000000153607.jpg'}, {'id': 329258, 'file_name': '00000032925 8.jpg'}, {'id': 452866, 'file_name': '000000452866.jpg'}]

subset of annotations [{'image_id': 542959, 'bbox': [32.52, 86.34, 8.53, 9.41], 'category_id': 62}, {'i mage_id': 542959, 'bbox': [98.12, 110.52, 1.95, 4.07], 'category_id': 86}, {'image_id': 542959, 'bbox': [91.28, 51.62, 3.95, 5.72], 'category_id': 86}, {'image_id': 542959, 'bbox': [110.48, 110.82, 14.55, 1 5.22], 'category_id': 62}, {'image_id': 542959, 'bbox': [96.63, 50.18, 18.67, 13.46], 'category_id': 6 2}]
```

Figure 6.21 – Contents of annotation dictionaries

6. The dictionaries that we created in the previous step aren't quite what we need to get complete annotations for the images. What we want is a consolidated annotation for each image that lists the bounding boxes for each object in the image along with the object's category. We could do this manually by manipulating the dictionaries we created in the previous step, but fastai provides a very handy function called `get_annotations` that does the work for us. Run the following cell to define the annotation structures:

```
image_files, bbox_lbl = get_annotations(path/'train.
json')
img_bbox_combo = dict(zip(image_files, bbox_lbl))
```

Here are the key parts of the code in this cell:

a) `aget_annotations(path/'train.json')`: Applies the `get_annotations` function to the `train.json` file to get an annotation structure. The output of this function is a list of filenames and a list of labeled bounding boxes.

b) `dict(zip(image_files, bbox_lbl))`: Creates a dictionary that combines the file list and the labeled bounding box list output from the previous command.

7. Run the following cell to examine one of the elements of the annotation dictionary, `img_bbox_combo`, that you created in the previous cell:

```
img_bbox_combo[image_files[5]]
```

The output of this cell, as shown in the following screenshot, shows that elements of the dictionary are tuples made up of a list of bounding boxes (sets of 2 x and y coordinates that define the top-left and bottom-right points of the box around the object) and a list of corresponding object categories:

```
([[103.73, 49.63, 125.98, 91.81],
  [18.25, 78.44, 36.980000000000004, 89.50999999999999]],
 ['tv', 'chair'])
```

Figure 6.22 – An element of img_bbox_combo

8. Run the following cell to see the image associated with the annotation you examined in the previous cell:

```
image_subpath = 'train/'+image_files[5]
img = PILImage.create(path/image_subpath)
img
```

The output of this cell is the following image:

Figure 6.23 – Example image

9. Run the following cell to define a function to return the bounding box associated with the input image file:

```
def get_bbox(filename):
    return np.array(img_bbox_combo[os.path.
basename(filename)][0])
```

10. Run the following cell to define a function to return the label (that is, the category) associated with the input image file:

```
def get_lbl(filename):
    return np.array(img_bbox_combo[os.path.
basename(filename)][1],dtype=object)
```

11. Run the following cell to define a function to return the image files in the dataset:

```
def get_items(noop):
    return get_image_files(path/'train')
```

12. Run the following cell to define a DataBlock object for the dataset using the functions that you defined in the previous three cells:

```
db = DataBlock(blocks=(ImageBlock, BBoxBlock,
BBoxLblBlock),
                get_items=get_image_files,
                splitter=RandomSplitter(),
                get_y=[get_bbox, get_lbl],
                n_inp=1)
```

For almost all of the recipes so far in this book, we have used the top-level fastai API, which means we would be defining some kind of dataloaders object at this step. For this dataset, however, we need more flexibility than we can get from a dataloaders object, which is why we have defined a DataBlock object here. For details on DataBlock objects, refer to the following documentation: https://docs.fast.ai/data.block.html#DataBlock.

Here are the arguments for the definition of the DataBlock object:

a) blocks=(ImageBlock, BBoxBlock, BBoxLblBlock): Specifies that the input to the model is images (ImageBlock) and the target has two parts: bounding boxes for object in the images (BBoxBlock), and labels (categories) associated with each of the bounding boxes (BBoxLblBlock).

b) get_items=get_image_files: Specifies that the get_image_files function is called to get the input to the DataBlock object.

c) get_y=[get_bbox, get_lbl]: Specifies the functions that are applied to the results of get_items. The image filenames are sent as arguments to each of these functions. The first function, get_bbox, returns the list of bounding boxes associated with the image file according to the annotation information we ingested from the train.json file. The second function, get_lbl, returns the list of labels (categories) for the bounding boxes associated with the image file, according to the annotation information we ingested from the train.json file.

d) n_inp=1: Specifies the number of elements in the tuples specified in the blocks argument that should be considered part of the input, in our case, just the image files.

13. Run the following cell to define a `dataloaders` object using the `DataBlock` object, `db`, you created in the previous cell:

```
dls = db.dataloaders(path,bs=32)
```

Here are the arguments for the `dataloaders` definition:

a) `path`: Specifies that the path to use for the `dataloaders` object is the `path` object you defined for the dataset at the beginning of the notebook

b) `bs=32`: Specifies that the batch size is 32

14. Now it is finally time to see what batches for the `dataloaders` object look like. Run the following cell to see a sample of entries from a batch:

```
dls.show_batch(max_n=4, figsize=(10,10))
```

Here are the arguments for `show_batch()`:

a) `max_n=4`: Specifies the number of samples to show

b) `figsize=(10,10)`: Specifies the dimensions of the samples

The output of `show_batch()` is shown in the following image. Note that you see the images with the labeled objects highlighted with bounding boxes and annotation text. Also note that you may see different sample images when you run `show_batch()` in your notebook:

Figure 6.24 – Sample image output by show_batch()

Congratulations! You have successfully ingested and prepared an image location dataset, one of the most complex data manipulation tasks that you will tackle in this book.

How it works...

After working through the recipe in this section, you may have some questions about the COCO_TINY dataset and the image location dataset in general. In this section, we will go through some of the questions that may have come up as you went through the recipe.

What kinds of objects are annotated in the images in the dataset?

If you rerun the following cell several times, you will see a variety of annotated images:

```
dls.show_batch(max_n=4, figsize=(10,10))
```

You can see an example of the output of this cell in *Figure 6.25*. Note that the annotations don't cover every possible object in the images. For example, the animal in the first image isn't annotated. Which objects are annotated for images in COCO_TINY? We'll answer that question in this section:

Figure 6.25 – Output of show_batch()

When you ran the following cell to bring the annotation file, `train.json`, into a Python dictionary, you created a sub-dictionary called `categories`, which contains all the categories of objects annotated in the COCO_TINY dataset as follows:

```
with open(path/'train.json') as json_file:
    data = json.load(json_file)
    # each nested structure is a list of dictionaries
    categories = data['categories']
    images = data['images']
    annotations = data['annotations']
```

To see what's in the `categories` dictionary, run the following cell:

```
print("categories ", categories)
```

The output of this cell is as follows:

```
categories  [{'id': 62, 'name': 'chair'}, {'id': 63, 'name': 'couch'}, {'id': 72, 'name': 'tv'}, {'id': 75, 'name':
'remote'}, {'id': 84, 'name': 'book'}, {'id': 86, 'name': 'vase'}]
```

Figure 6.26 – The categories dictionary

From this dictionary, you can see that there are only six objects annotated for the images in the COCO_TINY dataset: **chair, couch, tv, remote, book**, and **vase**.

If you run `show_batch()` repeatedly, you will see that even objects that are only partially shown in an image are annotated. For example, the first image in *Figure 6.25* has an annotation for a chair, as shown in the following image, even though only part of the legs of the chair is shown in the image:

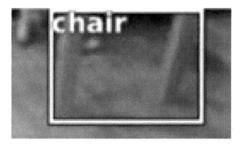

Figure 6.27 – Close-up of the chair annotation

You will also see that some images contain many annotated objects, and the bounding boxes for these objects can overlap, as shown in the following image:

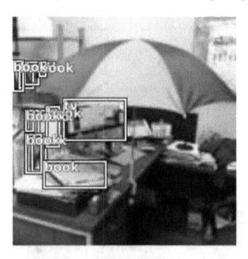

Figure 6.28 – An image with multiple, overlapping bounding boxes

This subsection describes which objects are annotated in the images in COCO_TINY. As you rerun the show_batch() command repeatedly, you will see for yourself how complex the annotations can be for some objects.

How are the bounding boxes for annotated objects defined?

While you were working through the recipe in this section, you may have asked yourself where the bounding boxes were defined. When you ran the following cell to bring the annotation file, train.json, into a Python dictionary, you created a sub-dictionary called annotations that contains all the annotations of objects annotated in the COCO_TINY dataset:

```
with open(path/'train.json') as json_file:
    data = json.load(json_file)
    # each nested structure is a list of dictionaries
    categories = data['categories']
    images = data['images']
    annotations = data['annotations']
```

You can run the following cell to see the contents of a subset of the annotations dictionary:

```
print("subset of annotations",list(annotations)[:5])
```

The output of this cell shows examples of annotations for objects in a particular image, as shown in the following screenshot:

```
subset of annotations [{'image_id': 542959, 'bbox': [32.52, 86.34, 8.53, 9.41], 'category_id': 62}, {'i
mage_id': 542959, 'bbox': [98.12, 110.52, 1.95, 4.07], 'category_id': 86}, {'image_id': 542959, 'bbox':
[91.28, 51.62, 3.95, 5.72], 'category_id': 86}, {'image_id': 542959, 'bbox': [110.48, 110.82, 14.55, 1
5.22], 'category_id': 62}, {'image_id': 542959, 'bbox': [96.63, 50.18, 18.67, 13.46], 'category_id': 6
2}]
```

Figure 6.29 – Example of annotations

The values for the bbox keys in this dictionary define the bounding boxes for objects by specifying the *x* and *y* values at the extremes of the objects.

Consider the following image. This image has a single annotated object – let's see how to view the coordinates of its bounding box:

Figure 6.30 – Example annotated image with a bounding box

The filename for this image is 000000071159.jpg. To see the bounding box for this image, run the following cell:

```
get_bbox(path/'train/000000071159.jpg')
```

The output for this cell shows the bounding box coordinates from this image. If the image had multiple objects annotated in it, there would be a bounding box defined for each object:

```
array([[13.52, 19.59, 44.  , 56.46]])
```

Figure 6.31 – Bounding box for the object in the example image in Figure 6.30

In this section, we have reviewed some of the details of the bounding box annotations in the COCO_TINY dataset.

Why didn't we train a model using COCO_TINY?

Unlike most of the recipes in this book, the recipe in this section did not include model training. After the effort we put into creating a dataloaders object for the COCO_TINY dataset, why didn't we go all the way and train a model with it?

The simple answer is that the fastai framework does not currently incorporate a simple way to train a model on an image location dataset such as COCO_TINY. If you want to attempt to train such a model, you can check out this repo for an approach, but be prepared to get into details well beyond the high-level fastai API that we have been exploring in this book so far: https://github.com/muellerzr/Practical-Deep-Learning-for-Coders-2.0/blob/master/Computer%20Vision/06_Object_Detection.ipynb.

Training a classification model with a standalone vision dataset

In the *Training a classification model with a simple curated vision dataset* recipe, you went through the steps to ingest a fastai curated dataset and use it to train an image classification model.

In this section, you will go through the same process for a standalone dataset called fruits-360. This dataset (described in more detail here: https://www.kaggle.com/moltean/fruits) contains over 90,000 images of fruits and vegetables organized into over 130 categories.

In this recipe, we'll begin by bringing this dataset into Gradient. Then we will work through the training_with_standalone_image_datasets.ipynb notebook to ingest the dataset and use it to train a fastai image classification model. Finally, we will see how well the trained model classifies images from the test set and save the trained model.

Getting ready

Ensure that you have uploaded the fruits-360 dataset to your Gradient environment by following these steps:

1. Download archive.zip from https://www.kaggle.com/moltean/fruits.

2. Upload `archive.zip` to your Gradient environment. You can use the upload button in JupyterLab in Gradient to do the upload, but you need to do it in several steps as follows:

    a) From the terminal in your Gradient environment, make `/notebooks` your current directory:

    cd /notebooks

    b) If you have not already created a `notebooks/temp` directory, make a new `/notebooks/temp` directory:

    mkdir temp

    c) In the JupyterLab file browser, make `temp` your current folder, select the upload button (see *Figure 6.32*), and select `archive.zip` from your local system folder where you downloaded it in *Step 1*:

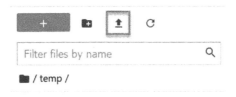

Figure 6.32 – Upload button in JupyterLab

3. Now that you have uploaded `archive.zip` to the `/notebooks/temp` directory in your Gradient environment, make that directory your current directory by running the following command in the Gradient terminal:

    cd /notebooks/temp

4. Unzip `archive.zip` into the `/storage/archive` directory by running the following command in the Gradient terminal:

    unzip archive.zip -d /storage/archive

5. Confirm that you now have the dataset set up in your Gradient environment by running the following command from within the Gradient terminal:

    cd /storage/archive/fruits-360

6. Then, run the following command to list the contents of the directory:

    ls

The output of this command should look like what is shown in the following screenshot:

```
LICENSE  Test  Training  papers  readme.md  test-multiple_fruits
```

Figure 6.33 – Contents of the fruits-360 directory

With these preparation steps, you have moved the files that make up the `fruits-360` dataset to the correct location in your Gradient environment to be used by a fastai model.

The `fruits-360` dataset featured in this section is introduced in the Kaggle competition **Fruits 360** from 2020 (`https://www.kaggle.com/moltean/fruits`). I am grateful for the opportunity to include an example using this dataset in this book.

> **Dataset citation**
> Mihai Oltean (2017-2020). *Fruits 360 – A dataset with 90,380 images of 131 fruits and vegetables* (`https://mihaioltean.github.io`).

How to do it...

In this section, you will be running through the `training_with_standalone_image_datasets.ipynb` notebook. Once you have the notebook open in your fastai environment, complete the following steps:

1. Run the cells in the notebook up to the `Ingest the dataset` cell to import the required libraries and set up your notebook.

2. Run the following cell to define the `path` object for this dataset:

```
path = URLs.path('fruits-360')
```

3. Run the following cell to examine the directory structure of the dataset:

```
path.ls()
```

The output shows the directory structure of the dataset, as shown in the following screenshot:

```
(#6) [Path('/storage/archive/fruits-360/Test'),Path('/storage/archive/fruits-360/papers'),Path('/storag
e/archive/fruits-360/LICENSE'),Path('/storage/archive/fruits-360/test-multiple_fruits'),Path('/storage/
archive/fruits-360/Training'),Path('/storage/archive/fruits-360/readme.md')]
```

Figure 6.34 – Output of path.ls()

4. Run the following cell to define an `ImageDataLoaders` object for this dataset:

```
dls = ImageDataLoaders.from_folder(path,
train='Training', valid='Test')
```

Here are the arguments for the definition of the `ImageDataLoaders` object:

a) `path`: Specifies that the `ImageDataLoaders` object is defined using the `path` object you created in the previous cell

b) `train='Training'`: Specifies that the training data is in the `/storage/archive/fruits-360/Training` directory

c) `valid='Test'`: Specifies that the validation data is in the `/storage/archive/fruits-360/Test` directory

5. Run the following cell to display a batch from the dataset:

```
dls.train.show_batch(max_n=4, nrows=1)
```

The output of this cell is a set of 4 items from a batch, showing images along with their corresponding categories, as shown in the following screenshot:

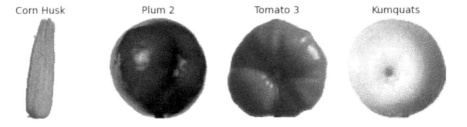

Figure 6.35 – Output of show_batch()

6. Run the following cell to examine the contents of the `Training` subdirectory:

```
(path/'Training').ls()
```

The output shows the structure of the `Training` subdirectory, as shown in the following screenshot:

```
(#131) [Path('/storage/archive/fruits-360/Training/Pomegranate'),Path('/storage/archive/fruits-360/Trai
ning/Pepper Green'),Path('/storage/archive/fruits-360/Training/Apple Golden 2'),Path('/storage/archive/
fruits-360/Training/Strawberry Wedge'),Path('/storage/archive/fruits-360/Training/Apple Crimson Snow'),
Path('/storage/archive/fruits-360/Training/Granadilla'),Path('/storage/archive/fruits-360/Training/Quin
ce'),Path('/storage/archive/fruits-360/Training/Tomato not Ripened'),Path('/storage/archive/fruits-360/
Training/Peach'),Path('/storage/archive/fruits-360/Training/Apricot')...]
```

Figure 6.36 – Contents of the Training subdirectory

7. Run the following cell to see a single item in the dataset:

```
img_files = get_image_files(path)
img = PILImage.create(img_files[100])
img
```

Here are the key elements of this cell:

a) `img_files = get_image_files(path)`: Specifies that `path` is recursively examined and returns all the image files in the path

b) `img = PILImage.create(img_files[100])`: Creates the image object, `img`, from a specific file returned by the previous statement

The output of this cell is one of the dataset files rendered as an image in the notebook:

Figure 6.37 – An image from the dataset

8. Run the following cell to define the model as a `cnn_learner` object:

```
learn = cnn_learner(dls, resnet18,
                    loss_
func=LabelSmoothingCrossEntropy(),
                    metrics=accuracy))
```

Here are the arguments for the definition of the `cnn_learner` object:

a) `dls`: Specifies that the `cnn_learner` object is defined using the `ImageDataLoaders` object you defined earlier in this notebook

b) `resnet18`: Specifies the pre-trained model to use as a starting point for this model

c) `loss_func=LabelSmoothingCrossEntropy()`: Specifies the loss function to use in the training process

d) `metrics=accuracy`: Specifies that `accuracy` is the performance metric that will be optimized in the training process

9. Run the following cell to train the model:

```
learn.fine_tune(5)
```

The argument indicates that the training run will be for 5 epochs.

The output displays the training loss, validation loss, and accuracy for each epoch, as shown in the following screenshot:

epoch	train_loss	valid_loss	accuracy	time
0	1.077759	1.049447	0.963946	01:31

epoch	train_loss	valid_loss	accuracy	time
0	0.913377	0.902737	0.991978	01:58
1	0.863309	0.864238	0.995769	01:58
2	0.841727	0.836321	0.997796	01:58
3	0.831393	0.830771	0.997444	01:58
4	0.829582	0.828693	0.997884	01:58

Figure 6.38 – Output of training the model

10. Let's try out the trained model on some examples from the test dataset. First, run the following cell to define an object for one of the images in the test dataset:

```
img_test_files = get_image_files(path/"Test")
img2 = PILImage.create(img_test_files[700])
img2
```

Here are the key elements of this cell:

a) `img_files = get_image_files(path/"Test"))`: Returns all the image files under the `Test` directory

b) `img = PILImage.create(img_files[700])`: Creates the image object, `img2`, from a specific file returned by the previous statement

The output of this cell is an image of a strawberry, as shown in the following screenshot:

Figure 6.39 – An image of a strawberry from the test dataset

11. Run the following cell to define an object for another one of the images in the test dataset:

```
img3 = PILImage.create(img_test_files[8000])
img3
```

The output of this cell is an image of a tomato, as shown in the following screenshot:

Figure 6.40 – An image of a tomato from the test dataset

12. Now that we have defined objects for a couple of images from the test dataset, let's exercise the trained image classification model on them. First, run the following cell to apply the model to the strawberry image:

```
learn.predict(img2)
```

The output of this cell is the model's prediction, as shown in the following screenshot. Note that the model correctly predicts the category of the image. The numbers shown in the `TensorImage` array correspond to the likelihood that the trained model ascribes to the image being in each of the categories:

```
('Strawberry Wedge',
 TensorImage(117),
 TensorImage([5.5287e-04, 5.8729e-04, 4.3915e-04, 7.1637e-04, 4.6952e-04, 5.7983e-04, 4.7180e-04, 5.961
8e-04, 5.0780e-04, 4.0931e-04, 4.5940e-04, 7.0770e-04, 7.0809e-04, 6.5259e-04, 6.5978e-04, 5.3266e-04,
       7.2990e-04, 8.4665e-04, 3.6720e-04, 7.0212e-04, 8.6347e-04, 6.0037e-04, 7.3940e-04, 7.6813e-0
4, 6.0635e-04, 7.5843e-04, 7.1088e-04, 6.3793e-04, 7.8211e-04, 5.6861e-04, 6.8963e-04, 6.6847e-04,
       6.0576e-04, 6.2287e-04, 7.5428e-04, 9.0836e-04, 6.8213e-04, 7.9148e-04, 5.5145e-04, 5.9907e-0
4, 5.0626e-04, 5.7467e-04, 6.3718e-04, 6.5144e-04, 5.0754e-04, 8.1586e-04, 5.3653e-04, 6.4462e-04,
       6.2153e-04, 6.6965e-04, 7.3855e-04, 7.9604e-04, 5.8874e-04, 5.4711e-04, 5.7097e-04, 6.0761e-0
4, 5.8197e-04, 7.9834e-04, 7.2468e-04, 6.7074e-04, 4.2638e-04, 6.4700e-04, 7.2001e-04, 4.7989e-04,
       6.2428e-04, 5.0561e-04, 8.9213e-04, 6.2297e-04, 6.2638e-04, 6.8443e-04, 6.0546e-04, 7.3738e-0
4, 5.9545e-04, 5.4710e-04, 5.5876e-04, 5.5428e-04, 7.0495e-04, 5.9575e-04, 6.4429e-04, 6.2548e-04,
       5.3481e-04, 7.3763e-04, 8.1653e-04, 6.7845e-04, 5.5469e-04, 5.8693e-04, 7.6158e-04, 5.2563e-0
4, 5.5775e-04, 8.9908e-04, 5.8551e-04, 7.3647e-04, 7.4814e-04, 6.9160e-04, 6.0768e-04, 6.0972e-04,
       6.8198e-04, 5.1440e-04, 5.4522e-04, 6.2201e-04, 6.9380e-04, 7.9181e-04, 7.0495e-04, 5.5718e-0
4, 8.1783e-04, 6.8215e-04, 5.9703e-04, 6.2589e-04, 6.8557e-04, 7.1313e-04, 7.0622e-04, 5.2479e-04,
       9.2585e-04, 6.6376e-04, 8.6684e-04, 1.4500e-03, 5.9743e-04, 9.1559e-01, 6.1752e-04, 5.8028e-0
4, 6.2902e-04, 6.2443e-04, 6.4723e-04, 5.7449e-04, 5.8301e-04, 6.6725e-04, 5.9484e-04, 6.9533e-04,
       5.5896e-04, 6.6607e-04, 5.7679e-04]))
```

Figure 6.41 – Image classification model's prediction on the image of the strawberry

13. Let's see how the model does on the tomato image. Run the following cell to apply the model to the image of the tomato:

```
learn.predict(img3)
```

The output of this cell is the model's prediction, as shown in the following screenshot. Note that the model correctly predicts that the image is a tomato:

```
('Tomato Heart',
 TensorImage(125),
 TensorImage([5.6110e-04, 7.4122e-04, 6.6106e-04, 7.0526e-04, 8.3456e-04, 7.1830e-04, 4.7857e-04, 8.251
0e-04, 7.0870e-04, 6.9801e-04, 6.7549e-04, 5.1823e-04, 7.5877e-04, 8.6114e-04, 6.3375e-04, 6.3768e-04,
       6.2907e-04, 7.3208e-04, 7.5443e-04, 6.9275e-04, 6.8134e-04, 5.4378e-04, 6.4222e-04, 6.1358e-0
4, 6.8001e-04, 6.4096e-04, 6.9659e-04, 8.1419e-04, 6.5037e-04, 8.4004e-04, 7.2150e-04, 6.6219e-04,
       9.5809e-04, 8.3602e-04, 7.3986e-04, 5.5498e-04, 6.7941e-04, 7.5781e-04, 5.0136e-04, 6.9421e-0
4, 7.4822e-04, 7.5638e-04, 6.0426e-04, 7.2764e-04, 8.8739e-04, 7.9243e-04, 5.5748e-04, 6.7485e-04,
       4.8520e-04, 6.9078e-04, 6.8147e-04, 6.5862e-04, 6.2541e-04, 6.0323e-04, 5.7093e-04, 6.2094e-0
4, 7.1028e-04, 6.9319e-04, 5.4292e-04, 7.9601e-04, 7.6126e-04, 7.2747e-04, 7.1577e-04, 7.7771e-04,
       5.9368e-04, 1.0395e-03, 7.4368e-04, 6.3575e-04, 6.7936e-04, 6.9432e-04, 7.1173e-04, 5.7752e-0
4, 1.0001e-03, 9.2254e-04, 1.0231e-03, 5.3364e-04, 7.3818e-04, 8.7064e-04, 6.2016e-04, 6.6928e-04,
       1.0332e-03, 7.8081e-04, 6.2899e-04, 6.0411e-04, 9.1333e-04, 5.1715e-04, 7.8582e-04, 5.9428e-0
4, 6.4144e-04, 5.9822e-04, 6.8863e-04, 7.1796e-04, 6.0904e-04, 6.8777e-04, 7.3537e-04, 9.1024e-04,
       8.6671e-04, 7.3775e-04, 6.8036e-04, 5.8841e-04, 5.7719e-04, 5.7521e-04, 5.9635e-04, 5.7538e-0
4, 8.4441e-04, 6.9781e-04, 6.4098e-04, 5.6899e-04, 1.0160e-03, 6.6534e-04, 4.9240e-04, 6.4860e-04,
       6.4336e-04, 5.6270e-04, 6.3225e-04, 4.6516e-04, 8.3983e-04, 9.1788e-04, 7.0737e-04, 5.5703e-0
4, 1.0298e-03, 7.0811e-04, 1.0442e-03, 5.1234e-04, 7.6787e-04, 9.0879e-01, 6.5822e-04, 5.9174e-04,
       6.9979e-04, 7.3096e-04, 7.3068e-04]))
```

Figure 6.42 – Image classification model's prediction on the image of a bird

14. The trained model seems to be doing a good job of predicting the category of images from the test set, but there is still some ambiguity. The fruit and vegetable images can be ambiguous to a human, so let's see how the trained model predicts the category of images where we know exactly what the category is. First, run the following cell to define an image of an avocado from the test dataset:

```
avocado_files = get_image_files(path/"Test/Avocado")
avocado_img = PILImage.create(avocado_files[30])
avocado_img
```

Here are the key elements of this cell:

a) `avocado_files = get_image_files(path/"Test/Avocado"))`: Returns all the image files under the `Test/Avocado` directory

b) `avocado_img = PILImage.create(avocado_files[30])`: Creates the image object, `avocado_img`, from a specific avocado image file from the file set returned by the previous statement

The output of this cell is an image of an avocado, as shown in the following screenshot:

Figure 6.43 – Image of an avocado from the test dataset

15. Let's get another image from a different directory in the test dataset. Run the following cell to define an image of a walnut from the test dataset:

```
walnut_files = get_image_files(path/"Test/Walnut")
walnut_img = PILImage.create(walnut_files[30])
walnut_img
```

Here are the key elements of this cell:

a) `walnut_files = get_image_files(path/"Test/Walnut")`: Returns all the image files under the `Test/Walnut` directory

b) `walnut_img = PILImage.create(walnut_files[30])`: Creates the image object, `walnut_img`, from a specific walnut image file from the file set returned by the previous statement

The output of this cell is an image of a walnut, as shown in the following screenshot:

Figure 6.44 – Image of a walnut from the test dataset

16. Now that we have defined objects for a couple of images from specific directories in the test dataset, let's exercise the trained image classification model on them. First, run the following cell to apply the model to the avocado image:

```
learn.predict(avocado_img)
```

The output of this cell is the model's prediction for the `avocado_img` image, as shown in the following screenshot. Note that the model correctly predicts the category of the image:

```
('Avocado',
 TensorImage(14),
 TensorImage([3.9023e-04, 5.9015e-04, 5.2299e-04, 3.0475e-04, 5.7779e-04, 5.7201e-04, 2.0197e-04, 3.021
1e-04, 5.0107e-04, 4.0196e-04, 5.1762e-04, 4.1946e-04, 3.3804e-04, 4.1061e-04, 9.3671e-01, 4.2566e-04,
        5.0626e-04, 5.1960e-04, 4.8361e-04, 4.6200e-04, 5.0016e-04, 3.9722e-04, 3.0203e-04, 5.4041e-0
4, 4.3439e-04, 5.5639e-04, 5.4460e-04, 4.9722e-04, 3.4571e-04, 4.7091e-04, 3.5859e-04, 5.6676e-04,
        5.3896e-04, 5.7201e-04, 7.3873e-04, 2.9858e-04, 3.9937e-04, 3.8848e-04, 5.1043e-04, 5.4287e-0
4, 5.5927e-04, 5.1005e-04, 4.3452e-04, 4.8029e-04, 5.6487e-04, 4.0782e-04, 3.7398e-04, 4.4078e-04,
        4.4194e-04, 6.0703e-04, 5.5859e-04, 3.9277e-04, 5.4818e-04, 5.0816e-04, 4.8134e-04, 5.8765e-0
4, 4.3756e-04, 3.7177e-04, 4.5130e-04, 6.5306e-04, 4.7650e-04, 7.4166e-04, 5.3843e-04, 3.6055e-04,
        5.1220e-04, 4.0618e-04, 3.5998e-04, 4.9659e-04, 7.1134e-04, 6.4496e-04, 3.9880e-04, 3.6079e-0
4, 5.4136e-04, 6.8846e-04, 5.4440e-04, 5.6530e-04, 5.4586e-04, 5.3762e-04, 7.0084e-04, 3.7813e-04,
        4.3333e-04, 5.8098e-04, 3.8209e-04, 5.5554e-04, 3.7787e-04, 6.4168e-04, 4.9021e-04, 5.2675e-0
4, 5.0481e-04, 4.8193e-04, 1.0131e-03, 5.8208e-04, 5.3925e-04, 5.1463e-04, 5.6995e-04, 3.5218e-04,
        5.6210e-04, 5.0859e-04, 6.1811e-04, 2.9358e-04, 4.6183e-04, 4.7107e-04, 6.3116e-04, 4.1540e-0
4, 4.7196e-04, 4.2009e-04, 4.6913e-04, 4.9917e-04, 3.3874e-04, 4.5602e-04, 2.4536e-04, 5.1899e-04,
        5.6236e-04, 3.8806e-04, 3.5534e-04, 5.8325e-04, 5.8201e-04, 5.3539e-04, 5.5095e-04, 4.0105e-0
4, 5.2433e-04, 4.7082e-04, 5.3796e-04, 4.5087e-04, 5.0284e-04, 4.0340e-04, 3.6922e-04, 5.2726e-04,
        4.8074e-04, 4.0830e-04, 4.5626e-04]))
```

Figure 6.45 – Image classification model's prediction on the image of the avocado

17. Let's see how the model does on the walnut image. Run the following cell to apply the model to the image of the walnut:

```
learn.predict(walnut_img)
```

The output of this cell is the model's prediction, as shown in the following screenshot. Note that the model correctly predicts that the image is a walnut:

```
('Walnut',
 TensorImage(129),
 TensorImage([5.0013e-04, 5.8467e-04, 7.0452e-04, 6.0763e-04, 5.5167e-04, 6.7749e-04, 6.8619e-04, 5.102
8e-04, 5.7429e-04, 5.2902e-04, 5.2863e-04, 6.0057e-04, 5.1680e-04, 5.1235e-04, 5.7513e-04, 5.0199e-04,
         6.1083e-04, 6.2190e-04, 6.0553e-04, 7.0514e-04, 5.3982e-04, 4.6631e-04, 4.9104e-04, 7.3324e-0
4, 6.3145e-04, 6.8161e-04, 5.5360e-04, 5.1731e-04, 6.0942e-04, 5.5742e-04, 6.3422e-04, 6.3248e-04,
         5.1715e-04, 5.9454e-04, 6.5074e-04, 4.4216e-04, 5.8860e-04, 6.0122e-04, 6.4125e-04, 5.6648e-0
4, 4.5822e-04, 6.1421e-04, 5.2089e-04, 5.8130e-04, 5.6028e-04, 6.7495e-04, 4.9246e-04, 4.8495e-04,
         6.5103e-04, 6.2217e-04, 5.3554e-04, 6.1928e-04, 6.9799e-04, 8.1177e-04, 5.2141e-04, 7.4727e-0
4, 6.9618e-04, 6.2065e-04, 6.5892e-04, 6.0229e-04, 6.7863e-04, 4.3354e-04, 5.3899e-04, 4.6352e-04,
         6.7922e-04, 5.6060e-04, 6.1658e-04, 6.1120e-04, 5.5102e-04, 6.2588e-04, 5.0936e-04, 5.3076e-0
4, 5.7705e-04, 6.3393e-04, 6.8977e-04, 5.1729e-04, 5.5937e-04, 6.2577e-04, 4.7848e-04, 6.4789e-04,
         5.0811e-04, 5.7817e-04, 5.9138e-04, 4.9482e-04, 6.5468e-04, 5.9456e-04, 6.5563e-04, 6.5299e-0
4, 6.5611e-04, 5.9734e-04, 5.9259e-04, 6.3149e-04, 5.2536e-04, 5.7820e-04, 5.4429e-04, 5.1889e-04,
         6.1921e-04, 5.5606e-04, 4.7429e-04, 4.8148e-04, 6.0877e-04, 6.3924e-04, 6.5298e-04, 5.9792e-0
4, 7.4838e-04, 6.2155e-04, 6.2797e-04, 7.1425e-04, 5.7054e-04, 8.5564e-04, 6.1362e-04, 7.3622e-04,
         4.8373e-04, 5.2082e-04, 5.8095e-04, 4.7680e-04, 6.1882e-04, 5.8547e-04, 5.6807e-04, 6.7309e-0
4, 5.5652e-04, 5.7659e-04, 5.8764e-04, 5.0513e-04, 5.6157e-04, 4.9809e-04, 5.2151e-04, 5.5212e-04,
         4.9715e-04, 9.2339e-01, 5.6719e-04]))
```

Figure 6.46 – Image classification model's prediction on the image of the walnut

18. Now that we have exercised the model to confirm that it makes good predictions for a small set of images, run the following cell to save the model:

```
learn.save("fruits_model"+modifier)
```

The output of this cell confirms that the model was saved in the `models` subdirectory of the dataset path, as shown in the following screenshot:

```
Path('/storage/archive/fruits-360/models/fruits_modelmay3.pth')
```

Figure 6.47 – Output of saving the model

Congratulations! You have trained a fastai image classification model on a large standalone image dataset and exercised the trained model with a set of examples from the dataset.

How it works...

If you compare the *Training a classification model with a simple curated vision dataset* recipe with the recipe in this section, you will notice that the code is very similar. In fact, once you have brought the `fruits-360` dataset into your Gradient environment, the code in the `training_with_standalone_image_datasets.ipynb` notebook is very close to the code in the `training_with_curated_image_datasets.ipynb` notebook. There are, however, some differences, as shown in the following list:

- The path definition statement is different. For the curated dataset, the definition of the path statement has a `URLs` object as its argument:

  ```
  path = untar_data(URLs.CIFAR)
  ```

  Whereas the path statement for the standalone dataset has the `fruits-360` directory as its argument:

  ```
  path = URLs.path('fruits-360')
  ```

- The test and train directories have different names between the two datasets.

- The standalone dataset is much bigger than the curated dataset, both in terms of the volume of images and the number of categories that the images are organized into.

- Out of the box, the trained model for the standalone dataset shows higher accuracy (over 95%) than the curated dataset (~80%).

Despite these differences, it is remarkable that fastai can train a model to categorize images in the `fruits-360` dataset so easily. As you saw in the recipe in this section, you were able to train the model on the `fruits-360` dataset and get great performance, with just a few lines of code. I think the model you created in this recipe is a great example of the power and flexibility of fastai.

Training a multi-image classification model with a curated vision dataset

In the *Training a classification model with a simple curated vision dataset* recipe, you went through the steps to ingest a fastai curated dataset and used it to train an image classification model.

In this section, you will go through the same process for another curated dataset called PASCAL_2007. This dataset (described in more detail here: `http://host.robots.ox.ac.uk/pascal/VOC/`) contains about 5,000 training images and the same number of test images. The dataset includes annotations that identify common objects that appear in each image. The identified objects are from 20 categories, including animals (cow, dog, cat, sheep, and horse), vehicles (boat, bus, train, airplane, bicycle, and car), and other items (person, sofa, bottle, and TV monitor).

The images in the CIFAR dataset introduced in the *Training a classification model with a simple curated vision dataset* recipe had a single labeled object. By contrast, the images in PASCAL_2007 can have zero, one, or more labeled objects. As you will see in this section, there are some challenges with training a model to predict multiple objects in the same image.

In this section, we will ingest the PASCAL_2007 dataset, including image annotations, explore the dataset, train a fastai model to categorize the images according to what objects are depicted in the images, and then exercise the trained model on some examples from the test dataset.

Getting ready

Confirm that you can open the `training_with_curated_multi_image_classification_datasets.ipynb` notebook in the ch6 directory of your repo.

The PASCAL_2007 dataset featured in this section is introduced in this paper – *The PASCAL Visual Object Classes (VOC) Challenge*: `http://host.robots.ox.ac.uk/pascal/VOC/pubs/everingham10.pdf`. I am grateful for the opportunity to include an example using this dataset in this book.

Dataset citation

Mark Everingham, Luc Van Gool, Christopher K. I. Williams, John Winn, and Andrew Zisserman (2008). *The PASCAL Visual Object Classes (VOC) Challenge* (`http://host.robots.ox.ac.uk/pascal/VOC/pubs/everingham10.pdf`).

How to do it...

In this section, you will be running through the `training_with_curated_multi_image_classification_datasets.ipynb` notebook. Once you have the notebook open in your fastai environment, complete the following steps:

1. Run the cells in the notebook up to the `Ingest the dataset` cell to import the required libraries and set up your notebook.

2. Run the following cell to define the `path` object for this dataset:

    ```
    path = untar_data(URLs.PASCAL_2007)
    ```

3. Run the following cell to examine the directory structure of the dataset:

    ```
    path.ls()
    ```

    The output shows the directory structure of the dataset, as shown in the following screenshot:

    ```
    (#8) [Path('/storage/data/pascal_2007/train'),Path('/storage/data/pascal_2007/test.json'),Path('/storage/data/pascal_2007/se
    gmentation'),Path('/storage/data/pascal_2007/train.json'),Path('/storage/data/pascal_2007/valid.json'),Path('/storage/data/p
    ascal_2007/test.csv'),Path('/storage/data/pascal_2007/train.csv'),Path('/storage/data/pascal_2007/test')]
    ```

    Figure 6.48 – Output of path.ls()

    Let's now review the key items in the directory structure of the `PASCAL_2007` dataset that we will be using in this recipe:

    a) `/storage/data/pascal_2007/train`: A directory containing images for training the model

    b) `/storage/data/pascal_2007/test`: A directory containing images for testing the trained model

    c) `/storage/data/pascal_2007/train.json`: A file containing annotations for the training images

    d) `/storage/data/pascal_2007/test.json`: A file containing annotations for test images

4. Run the following cell to bring the annotations in the `train.json` file into Python objects:

```
with open(path/'train.json') as json_file:
    data = json.load(json_file)
    # each nested structure is a list of dictionaries
    categories = data['categories']
    images = data['images']
    annotations = data['annotations']
```

Here are the key parts of the code used in this cell:

a) `data = json.load(json_file)`: Loads the contents of the whole `train.json` file into the `data` dictionary.

b) `categories = data['categories']`: Creates a separate list of dictionaries just for the category definitions. This dictionary defines the objects in the images in the dataset.

c) `images = data['images']`: Creates a separate list of dictionaries just for image files.

d) `annotations = data['annotations']`: Creates a separate list of dictionaries just for the annotations that specify the objects in the images and their bounding boxes.

As you will see in later steps in this recipe, we will not be using the `categories`, `images`, and `annotations` lists of dictionaries to extract the annotations to feed into the fastai model training process. Instead, we will be using the fastai built-in API, `get_annotations`, to work directly with the annotations. The dictionaries we define in this cell will, however, be useful to help us to understand the details regarding the annotations.

5. Run the following cell to see some examples of annotations:

```
print("categories ", categories)
print()
print("subset of images",list(images)[:5])
print()
print("subset of annotations",list(annotations)[:5])
```

The output of this cell lists entries from each of the lists of dictionaries that you created in the previous cell. As shown in the following screenshot, the `categories` list of dictionaries contains the categories of objects that can appear in the images along with their IDs:

```
categories [{'supercategory': 'none', 'id': 1, 'name': 'aeroplane'}, {'supercategory': 'none', 'id': 2, 'name':
'bicycle'}, {'supercategory': 'none', 'id': 3, 'name': 'bird'}, {'supercategory': 'none', 'id': 4, 'name': 'boa
t'}, {'supercategory': 'none', 'id': 5, 'name': 'bottle'}, {'supercategory': 'none', 'id': 6, 'name': 'bus'},
{'supercategory': 'none', 'id': 7, 'name': 'car'}, {'supercategory': 'none', 'id': 8, 'name': 'cat'}, {'supercat
egory': 'none', 'id': 9, 'name': 'chair'}, {'supercategory': 'none', 'id': 10, 'name': 'cow'}, {'supercategory':
'none', 'id': 11, 'name': 'diningtable'}, {'supercategory': 'none', 'id': 12, 'name': 'dog'}, {'supercategory':
'none', 'id': 13, 'name': 'horse'}, {'supercategory': 'none', 'id': 14, 'name': 'motorbike'}, {'supercategory':
'none', 'id': 15, 'name': 'person'}, {'supercategory': 'none', 'id': 16, 'name': 'pottedplant'}, {'supercategor
y': 'none', 'id': 17, 'name': 'sheep'}, {'supercategory': 'none', 'id': 18, 'name': 'sofa'}, {'supercategory':
'none', 'id': 19, 'name': 'train'}, {'supercategory': 'none', 'id': 20, 'name': 'tvmonitor'}]
```

Figure 6.49 – Entries from the categories list of dictionaries

The following screenshot shows that the `images` list of dictionaries lists the filenames of the image files in the training set along with their IDs and dimensions:

```
subset of images [{'file_name': '000012.jpg', 'height': 333, 'width': 500, 'id': 12}, {'file_name': '000017.jp
g', 'height': 364, 'width': 480, 'id': 17}, {'file_name': '000023.jpg', 'height': 500, 'width': 334, 'id': 23},
{'file_name': '000026.jpg', 'height': 333, 'width': 500, 'id': 26}, {'file_name': '000032.jpg', 'height': 281,
'width': 500, 'id': 32}]
```

Figure 6.50 – Entries from the images list of dictionaries

The following screenshot shows that `annotations` lists the bounding boxes and category IDs of the objects that appear in the image with the given `image_id`:

```
subset of annotations [{'segmentation': [[155, 96, 155, 270, 351, 270, 351, 96]], 'area': 34104, 'iscrowd': 0,
'image_id': 12, 'bbox': [155, 96, 196, 174], 'category_id': 7, 'id': 1, 'ignore': 0}, {'segmentation': [[184, 6
1, 184, 199, 279, 199, 279, 61]], 'area': 13110, 'iscrowd': 0, 'image_id': 17, 'bbox': [184, 61, 95, 138], 'cate
gory_id': 15, 'id': 2, 'ignore': 0}, {'segmentation': [[89, 77, 89, 336, 403, 336, 403, 77]], 'area': 81326, 'is
crowd': 0, 'image_id': 17, 'bbox': [89, 77, 314, 259], 'category_id': 13, 'id': 3, 'ignore': 0}, {'segmentatio
n': [[8, 229, 8, 500, 245, 500, 245, 229]], 'area': 64227, 'iscrowd': 0, 'image_id': 23, 'bbox': [8, 229, 237, 2
71], 'category_id': 2, 'id': 4, 'ignore': 0}, {'segmentation': [[229, 219, 229, 500, 334, 500, 334, 219]], 'are
a': 29505, 'iscrowd': 0, 'image_id': 23, 'bbox': [229, 219, 105, 281], 'category_id': 2, 'id': 5, 'ignore': 0}]
```

Figure 6.51 – Entries from the annotations list of dictionaries

6. We will use the fastai function, `get_annotations`, to get all the annotation information that we need for a given image, that is, the categories for the objects that are depicted in the image. Run the following cell to define the required annotation structures:

```
image_files, bbox_lbl = get_annotations(path/'train.
json')
img_bbox_combo = dict(zip(image_files, bbox_lbl))
```

Here are the key parts of the code used in this cell:

a) `get_annotations(path/'train.json')`: Applies the `get_annotations` function to the `train.json` file to get an annotation structure. The output of this function is a list of filenames and a list containing bounding boxes and object categories for each object in the image. The following screenshot shows an example of the contents of `bbox_lbl`, specifying the bounding boxes and categories for three objects:

```
([[8, 106, 499, 263], [420, 199, 482, 226], [324, 187, 411, 223]],
 ['aeroplane', 'aeroplane', 'aeroplane'])
```

Figure 6.52 – Example bounding box and categories

b) `dict(zip(image_files, bbox_lbl))`: Creates a dictionary that combines the file list and the labeled bounding box list output from the previous command.

7. Let's now take a look at one of the image files from the training set along with its annotations. First, run the following cell to see the annotations for a specific image file:

```
img_bbox_combo[image_files[5]]
```

The output shows the annotations associated with this image file:

```
([[8, 106, 499, 263], [420, 199, 482, 226], [324, 187, 411, 223]],
 ['aeroplane', 'aeroplane', 'aeroplane'])
```

Figure 6.53 – Annotations for an image file

8. Run the following cell to take a look at the image whose annotations we examined in the previous cell:

```
image_subpath = 'train/'+image_files[5]
img = PILImage.create(path/image_subpath)
img
```

The output shows the image associated with image_files[5]. As you can see in the following image, this image does indeed contain three planes, as indicated in the annotations you saw in the previous cell:

Figure 6.54 – Image whose annotation indicates it contains three planes

9. Run the following cell to define the get_category function to get values out of the list of dictionaries you created earlier. We will use this function later to examine the annotations of images in the test set:

```
def get_category(in_key_value,in_key,out_key,dict_list):
    return([cat[out_key] for cat in dict_list if cat[in_
key]==in_key_value] )
```

10. Run the following cell to define the get_lbl function that takes a filename as input and returns the list of category names from the annotation structure for that file:

```
def get_lbl(filename):
    return np.array(img_bbox_combo[os.path.
basename(filename)][1],dtype=object)
```

Here are the key parts of this function:

a) os.path.basename(filename): Returns the final part of the fully qualified filename. For example, if filename is the fully qualified name, /storage/ data/pascal_2007/train/006635.jpg, os.path.basename returns 006635.jpg. This conversion is required because the input image files will include fully qualified paths, but the img_bbox_combo structure is indexed with just the final part of the filename.

b) img_bbox_combo[os.path.basename(filename)][1]: Returns the categories associated with the filename image file.

11. Run the following cell to see an example of how get_lbl works:

```
get_lbl('/storage/data/pascal_2007/train/007911.jpg')
```

The output, as shown in the following screenshot, is a NumPy array containing the categories associated with the /storage/data/pascal_2007/ train/006635.jpg image file:

```
array(['motorbike', 'person', 'person', 'person', 'person', 'person', 'person', 'person', 'person', 'person', 'person', 'per
son', 'person', 'person'], dtype=object)
```

Figure 6.55 – Sample output of get_lbl

12. It looks like a lot is going on in the image we used as an argument to get_lbl in the previous cell. Run the following cell to take a look at the image:

```
image_subpath = 'train/007911.jpg'
img = PILImage.create(path/image_subpath)
img
```

As shown in the following image, this image matches the annotation we saw in the previous cell. It does indeed contain a motorbike along with a bunch of people:

Figure 6.56 – Image with a motorbike and several people

13. Run the following cell to examine the number of files in the training and test sets for this dataset:

```
print("number of training images: ",len(get_image_
files(path/'train')))
print("number of testing images: ",len(get_image_
files(path/'test')))
```

The output, shown in in the following screenshot, shows the number of files in each set:

```
number of training images:  5011
number of testing images:   4952
```

Figure 6.57 – Count of files in the training and test sets

14. Run the following cells to see the number of categories in the dataset:

```
print("number of categories is: ",len(categories))
```

The output, shown in the following screenshot, shows the number of categories in the dataset:

```
number of categories is:  20
```

Figure 6.58 – Number of categories in the dataset

15. Run the following cell to define the get_items function that will be used in the DataBlock definition to get the input data files:

```
def get_items(noop):
    return_list = []
    empty_list = []
    # filter the training files and keep only the ones
with valid info in the JSON file
    for file_path in get_image_files(path/'train'):
        file_id_list = get_category(os.path.
basename(file_path),'file_name','id',images)
        if len(file_id_list) > 0:
            return_list.append(file_path)
        else:
            empty_list.append(file_path)
    print("len(return_list): ",len(return_list))
    print("len(empty_list): ",len(empty_list))
    return(return_list)
```

Here are the key items in this function's definition:

a) return_list = []: Initializes the list of image files in the training set that have annotations.

b) empty_list = []: Initializes the list of image files in the training set that do not have annotations.

c) for file_path in get_image_files(path/'train'): Iterates through the files in the training set.

d) `file_id_list`: Is the list of file IDs in the annotation corresponding with the current `file_path`.

e) `if len(file_id_list) > 0`: Checks to see whether `file_id_list` has any entries. If it does, the function appends the current `file_path` to `return_list`. Otherwise, the function appends the current `file_path` to `empty_list`.

f) `return(return_list)`: This only returns the subset of image files that have annotations associated with them. If you include any of the image files that have no annotations associated with them, you will get an error in the step where you define a `dataloaders` object.

16. Run the following cell to define a `DataBlock` object for the dataset:

```
db = DataBlock(blocks=(ImageBlock, MultiCategoryBlock),
               get_items = get_items,
               splitter=RandomSplitter(),
               get_y=[get_lbl],
               item_tfms = RandomResizedCrop(128,\
   min_scale=0.35),
               n_inp=1)
```

Here are the key items in the definition of the `DataBlock` object:

a) `blocks=(ImageBlock, MultiCategoryBlock)`: Specifies the type of the input data (image files) and that the dataset has multi-category labels, that is, each image can have multiple annotated objects in it.

b) `get_items = get_items`: Specifies the function to apply to get the input items; in this case, the `get_items` function we defined in the previous cell that returns all the files in the training set that have annotations associated with them.

c) `splitter=RandomSplitter()`: Tells fastai to create a validation set from randomly selected items from the training set, by default using 20% of the items in the training set.

d) get_y=[get_lbl]: Specifies the function to get labels for the input, in this case, the get_lbl function. This function takes a filename as input and returns the list of categories from the annotations for that file.

e) item_tfms = RandomResizedCrop(168, min_scale=0.3): Specifies a transformation to apply during training. Because the image files in the training set are different sizes, we need to transform them all to a common size or we will get errors in show_batch. This transformation resizes the images by cropping them so they are all a common size.

f) n_inp=1: Specifies which of the elements defined in the blocks clause of the definition should be considered inputs, in this case, 1 or just ImageBlock.

17. Run the following cell to define a dataloaders object using the DataBlock object, db, that you created in the previous cell:

```
dls = db.dataloaders(path,bs=32)
```

Here are the arguments to the dataloaders definition:

a) path: Specifies that the source for the dataloaders object is the path object you created earlier in the notebook

b) bs=32: Specifies that the batch size is 32

The output of this cell, as shown in the following screenshot, shows the count of elements in return_list (the list of image files with valid annotations) and empty_list (the list of image files without valid annotations):

```
len(return_list):  2501
len(empty_list):  2510
```

Figure 6.59 – Output of the dataloader definition

18. Run the following cell to display a set of samples from a batch:

```
dls.show_batch(max_n=4, figsize=(10,10))
```

The output shows a selection of images from the training set along with the categories for the objects in the images that are described in the annotation corresponding to the image files, as shown in the following screenshot:

Figure 6.60 – Results of show_batch

19. Run the following cell to define the model by specifying a cnn_learner object:

```
learn = cnn_learner(dls, resnet18)
```

Here are the arguments to the cnn_learner definition:

a) dls: Specifies that the model is trained using the dataloaders object you defined in the previous cell

b) resnet18: Specifies that the model is based on the pre-trained resnet18 mode.

20. Run the following cell to train the model for 10 epochs:

```
learn.fine_tune(10)
```

The output lists the training and validation loss for each epoch, as shown in the following screenshot. The performance of the model improves as the validation loss decreases:

epoch	train_loss	valid_loss	time
0	0.877005	0.559717	00:07

epoch	train_loss	valid_loss	time
0	0.642405	0.471323	00:09
1	0.523728	0.292567	00:09
2	0.340163	0.150779	00:09
3	0.216213	0.119942	00:09
4	0.158638	0.109708	00:09
5	0.129925	0.107407	00:09
6	0.112691	0.105818	00:09
7	0.102112	0.103462	00:09
8	0.094495	0.103746	00:09
9	0.091821	0.103420	00:09

Figure 6.61 – Output of training the multi-category image classification model

21. Now that we have trained the model, let's exercise it on some images from the test set. To start with, run the following cell to prepare and display one of the images from the test set:

```
img_test_files = get_image_files(path/"test")
img2 = PILImage.create(img_test_files[100])
img2
```

This cell displays the following image:

Figure 6.62 – Image from the test set

22. Run the following cell to apply the trained model to the image displayed in the previous cell:

```
learn.predict(img2)
```

From the output, as shown in the following screenshot, you can see that the model gets the right category for one of the objects in the image, but is wrong about the second object in the image:

```
((#2) ['horse','person'],
 TensorImage([False, False, False, False, False, False, False, False, False, False, False, False,  True, F
alse,  True, False, False, False, False, False]),
 TensorImage([0.0020, 0.0067, 0.0016, 0.0090, 0.0063, 0.0034, 0.0185, 0.0066, 0.0041, 0.0366, 0.0059, 0.16
36, 0.7742, 0.0050, 0.9009, 0.0044, 0.0075, 0.0075, 0.0099, 0.0031]))
```

Figure 6.63 – Model prediction on the image of the horse

23. Run the following cell to prepare and display another image from the test set:

```
img3 = PILImage.create(img_test_files[200])
img3
```

This cell displays the following image:

Figure 6.64 – Image from the test set

24. Run the following cell to apply the trained model to the image displayed in the previous cell:

```
learn.predict(img3)
```

From the output, as shown in the following screenshot, you can see that the model gets the correct category for the object in the image:

```
((#1) ['cat'],
 TensorImage([False, False, False, False, False, False, False,  True, False, False, False, False, False, F
alse, False, False, False, False, False]),
 TensorImage([0.0012, 0.0024, 0.0167, 0.0028, 0.0041, 0.0028, 0.0018, 0.6394, 0.0068, 0.0207, 0.0027, 0.08
74, 0.0098, 0.0033, 0.0026, 0.0027, 0.0050, 0.0075, 0.0026, 0.0045]))
```

Figure 6.65 – Model prediction on the image of the cat

Congratulations! You have trained a fastai model that categorizes multiple objects in images.

How it works...

The recipe in this section is one of the longest, most complex recipes in this book. The following are some aspects of the recipe that may not have been evident to you as you worked through the steps.

This recipe uses transfer learning

If you look at the code from *Step 19* and *Step 20* of the recipe, you will see a pattern similar to the model definition and training steps from the recipes in the *Training a classification model with a simple curated vision dataset* recipe and the *Training a classification model with a standalone vision dataset* recipe.

In all of these recipes, you specified a pre-trained model in the model definition and then trained the model using a `fine_tune` statement. Taking advantage of the pre-trained vision classification model for this recipe makes a lot of sense because we had a fairly small training set, as you will see in the following section.

This recipe has a small training set for a deep learning model

How big was the training set for this recipe? The initial training set is a little over 5,000 images, as you can see in the count of the training set from *Step 13* of the recipe:

```
number of training images:  5011
number of testing images:   4952
```

Figure 6.66 – Count of items in the training and test sets

However, in *Step 15*, where we defined the `get_items` function, we had to filter the list of training images to only include those that had valid category annotations. As you can see from the output of *Step 17*, where we define the `dataloaders` object and invoke the `get_items` function, less than half of the input training images have valid annotations:

```
len(return_list):  2501
len(empty_list):   2510
```

Figure 6.67 – Count of items in return_list and empty_list

What does this mean? It means that we have 2,500 images to train our model. Having only 2,500 images to train a complex model has consequences, as we will see in the next section.

The trained model from this recipe does not have great performance

The model we trained in the recipe in this section does not have outstanding performance. It usually identifies one object in an image correctly, but it often doesn't identify more than one object, and when it does identify a second or third object, it often assigns these objects to an incorrect category. Why is the performance of this model not so good?

In this recipe, we used only 2,500 images to train a complex model to categorize multiple objects in images into 20 categories. Compare this to the `fruits-360` dataset used in the *Training a classification model with a standalone vision dataset* recipe. The `fruits-360` dataset has over 90,000 images. To see how many files there are in the training set, run the following command in your Gradient environment:

```
find /storage/archive/fruits-360/Training -type f | wc -l
```

The following screenshot shows the output of this command:

```
root@nyyka0v6hr:/storage/archive/fruits-360# find /storage/archive/fruits-360/Training -type f | wc -l
67692
```

Figure 6.68 – Count of images in the fruits-360 training set

There are over 67,500 files in the training set for the `fruits-360` dataset. This means that for the first curated image dataset problem we saw in this chapter, we had 25 times as many training samples as we had for the recipe in this section, and in the first instance, we applied the model to the simpler problem of identifying a single object in each image. The relative lack of training images in the `PASCAL_2007` dataset could partially explain the mediocre performance of the model we trained with this dataset.

Test your knowledge

Of the four application areas that fastai explicitly supports (tabular, text, recommender systems, and image/vision), fastai provides the most thorough support for creating models that work with image datasets. In this chapter, we have just scratched the surface of what you can do with fastai and image datasets. In this section, you will get a chance to dig a bit deeper into one of the fastai image dataset recipes from this chapter.

Getting ready

Ensure that you have followed the *Training a multi-image classification model with a curated vision dataset* recipe. In this section, you will be adapting the notebook you worked through in that recipe to try some new variations on deep learning with image datasets.

How to do it...

You can follow the steps in this section to try some variations on the image classification model that you trained with the PASCAL_2007 dataset in the *Training a multi-image classification model with a curated vision dataset* recipe:

1. Make a copy of the `training_with_curated_multi_image_classification_datasets.ipynb` notebook that you worked through in the *Training a multi-image classification model with a curated vision dataset* recipe. Give your new copy of the notebook the following name: `training_with_curated_multi_image_classification_datasets_variations.ipynb`.

2. Run the notebook up to and including the model definition cell:

```
learn = cnn_learner(dls, resnet18)
```

3. Add the following cell immediately after this cell and run it:

```
learn.summary()
```

The output of this cell lists the structure of the model. Note the summary of trainable and non-trainable parameters at the end of the output, as shown in the following screenshot:

```
Total params: 11,714,112
Total trainable params: 547,200
Total non-trainable params: 11,166,912
```

Figure 6.69 – Trainable parameter description at the end of the summary output before fine-tuning

4. Update the model training cell to train for 20 epochs and then run it:

```
learn.fine_tune(20)
```

The output of this cell, as shown in the following screenshot, shows how the training loss and validation loss develop through the epochs. Note how the training loss decreases steadily right up to the 20th epoch, while the validation loss decreases up to the 10th epoch, after which it goes up and down. What problem is evident in a situation like this, where the training loss keeps dropping but the validation loss stops dropping?

epoch	train_loss	valid_loss	time
0	0.883540	0.556862	00:08

epoch	train_loss	valid_loss	time
0	0.640981	0.464962	00:10
1	0.563007	0.366122	00:10
2	0.442997	0.236964	00:10
3	0.308103	0.149992	00:10
4	0.212907	0.120228	00:10
5	0.163879	0.111883	00:10
6	0.135566	0.107638	00:11
7	0.115231	0.109443	00:10
8	0.100900	0.104464	00:10
9	0.089603	0.103160	00:10
10	0.082206	0.105580	00:10
11	0.074104	0.105146	00:10
12	0.065566	0.103051	00:10
13	0.058820	0.106906	00:10
14	0.055226	0.104793	00:10
15	0.050850	0.106087	00:10
16	0.047641	0.105736	00:10
17	0.044487	0.106318	00:10
18	0.043247	0.105515	00:10
19	0.042669	0.105577	00:10

Figure 6.70 – Results of a 20-epoch training run

5. The previous step shows us that training for more epochs won't improve the model's performance. You can validate this for yourself by running the rest of the notebook and comparing how well the model trained with 20 epochs predicts the objects in the selected test images.

6. Add the following immediately after the model training cell and run it:

```
learn.summary()
```

The output of this cell lists the structure of the model. Note the summary of trainable and non-trainable parameters at the end of the output, as shown in the following screenshot. Compare this output to the output from summary() prior to fine-tuning the model, as shown in *Step 3*. What has changed to explain the difference in the output of summary() between *Step 3* and this step?

```
Total params: 11,714,112
Total trainable params: 11,714,112
Total non-trainable params: 0
```

Figure 6.71 – Trainable parameter description at the end of the summary output after fine-tuning

7. In the *How it works...* section of the previous recipe, we looked at the size of the training set for PASCAL_2007 and suggested that its small size could be a reason for the unimpressive performance of the model. One other potential culprit is the way that we are resizing the images in the definition of the DataBlock object, shown in the following cell:

```
db = DataBlock(blocks=(ImageBlock, MultiCategoryBlock),
               get_items = get_items,
               splitter=RandomSplitter(),
               get_y=[get_lbl],
               item_tfms = RandomResizedCrop(168,\
    min_scale=0.3),
               n_inp=1)
```

The transformation that resizes the images is RandomResizedCrop(168, min_scale=0.3). Update the DataBlock definition cell to try different image transformations. First, update the RandomResizedCrop function call to try a different image size and a different min_scale value. Train and exercise the model again on samples from the test set to see whether the performance changes for different image sizes and min_scale values.

8. Try using `Resize` rather than `RandomResizedCrop` as the transformation function (as shown in the following cell) and train the model again. See whether you get better results when you exercise the trained model on sample images from the training set:

```
db = DataBlock(blocks=(ImageBlock, MultiCategoryBlock),
               get_items = get_items,
               splitter=RandomSplitter(),
               get_y=[get_lbl],
               item_tfms = Resize(168),
               n_inp=1)
```

Congratulations! You have completed a review of training the fastai model on image datasets.

7
Deployment and Model Maintenance

So far in this book, you have trained a wide variety of fastai models, including models trained with tabular datasets, models trained with text datasets, recommender systems, and models trained with image data. All the models that you have trained have been exercised in the context of Jupyter notebooks. Jupyter notebooks are great for training models and exercising them with a couple of test examples, but what about actually making your model useful? How do you make your model available to other people or applications to actually solve problems?

The process of making your deep learning models available to other people or applications is called **deployment**. In this chapter, we will go through recipes that show how to deploy your fastai models. The industrial-strength production deployment of deep learning models is beyond the scope of this book. Instead, in this chapter, you will learn how to create simple, self-contained deployments that you can serve from your own local system.

Here are the recipes that will be covered in this chapter:

- Setting up fastai on your local system
- Deploying a fastai model trained on a tabular dataset
- Deploying a fastai model trained on an image dataset

- Maintaining your fastai model
- Test your knowledge

Technical requirements

In this chapter, you will be running deployments on your local system, which requires having fastai installed on your local system. To run fastai locally, a Windows or Linux system is recommended, with Python installed. While fastai can be installed on macOS, you will save yourself a lot of headaches if you use a Windows or Linux system for your local installation of fastai.

Ensure that you have cloned the repo for the book at `https://github.com/PacktPublishing/Deep-Learning-with-fastai-Cookbook` and have access to the `ch7` folder. This folder contains the code samples described in this chapter.

Setting up fastai on your local system

The first step in being able to do a simple web deployment of a fastai deep learning model is to set up your local system with PyTorch and fastai. You need to do this because you will be running code on your local system that invokes models that you trained earlier in this book. To exercise models to make predictions on your local system, you need to have the fastai framework installed. In this recipe, you will see how to set up fastai on your local system and how to validate your installation.

Getting ready

Ensure that you have Python (at least 3.7) installed on your local system.

To check the level of Python, enter the following command on the command line:

```
python –version
```

The output will show the version of Python installed on your local system as follows:

```
Python 3.7.4
```

Figure 7.1 – Python version

Ensure that you have cloned the book's repo at `https://github.com/PacktPublishing/Deep-Learning-with-fastai-Cookbook` to your local system.

How to do it...

To set up fastai on your local system, you need to set up PyTorch (the deep learning framework upon which fastai runs) and then fastai. To do this, run through the following steps:

1. Install PyTorch on your local system by running the following command in a terminal or command window of your local system. You can find complete details about installing PyTorch on your local system here: `https://pytorch.org/ get-started/locally/`:

    ```
    pip3 install torch==1.8.1+cpu torchvision==0.9.1+cpu
    torchaudio===0.8.1 -f https://download.pytorch.org/whl/
    torch_stable.html
    ```

2. Install fastai on your local system by following the instructions here for your operating system and typical Python installation approach: `https://docs. fast.ai/`.

3. Once you have installed PyTorch and fastai, validate your installation by opening up the `validate_local_setup.ipynb` notebook from the `ch7` directory in your local repo and run the following cell:

    ```
    import fastai
    fastai.__version__
    ```

Congratulations! You have successfully set up fastai on your local system.

How it works...

You may be asking why it's necessary to set up fastai on a local system to demonstrate how to deploy a fastai model. While it is possible to deploy fastai models without using your local system, there are several advantages to installing fastai locally:

1. You have complete control over the entire environment. By installing fastai locally, you can control the entire stack, from the level of pandas to the details of the web pages that you will use for deployment.

2. By deploying fastai models locally, you will avoid shortcuts that could limit your complete appreciation of how fastai models work when they are deployed. The deployments that you will see in this chapter may be simple but they are complete. By working through recipes where nothing is left as a black box, you will gain a thorough understanding of what is really happening when a fastai model is deployed.

3. If you are serious about exploiting fastai, it is handy to have a local installation. Back in *Chapter 1, Getting Started with fastai*, I specified that you would need a cloud environment, either Gradient or Colab, to run through the recipes in this book. Most fastai applications require GPUs to be trained efficiently. Setting up a GPU on an existing local system is not easy, and buying a pre-configured system with a GPU only makes sense if you are fully committed to exploiting the GPU regularly by working full-time on deep learning applications. So, a cloud environment that is GPU-enabled is the best place to start. However, having a working fastai environment on your local system can be very useful even if you're not going to be using the local system for model training. For example, in the process of writing this book, there were a couple of instances where there was a problem with the Gradient environment where I did most of the development of recipes for this book. Because I have fastai installed locally, when Gradient was unavailable, I could still use my local system to make progress on coding outside of the model.

4. If you don't already have some exposure to web application development, you will benefit from the brief experience you will get in this chapter. In my experience, many data scientists have zero idea of how web applications work, and since most of what we work on will eventually be rendered in one form or another in a web framework, it behooves us to get a basic understanding of how web applications work. By using a combination of the Python Flask library and basic HTML and JavaScript, we will create a very simple, yet complete, web application that illustrates some basic web application principles. If you haven't come across these principles before, you will find them useful to have in your toolbox.

I hope that this background helps to convince you of the value of having a working fastai setup on your local system. Now that you have completed the fastai setup, you are ready for the subsequent sections where you deploy models on your local system.

Deploying a fastai model trained on a tabular dataset

Back in the *Saving a trained tabular model* recipe in *Chapter 3, Training Models with Tabular Data*, you exercised a fastai model that you had saved. Recall the steps you went through in the recipe.

First, you loaded the saved model as follows:

```
learn = load_learner('/storage/data/adult_sample/adult_sample_
model.pkl')
```

Then you took a test sample and generated a prediction from the model for the test sample:

```
test_sample = df_test.iloc[0]
learn.predict(test_sample)
```

The output of the prediction, as shown in the following screenshot, included the values of the input sample, the prediction, and probability of each outcome for the prediction:

```
(   workclass  education  marital-status  occupation  relationship  race  sex  \
 0        5.0        8.0             3.0         0.0           6.0   5.0  1.0

    native-country  education-num_na  age     fnlwgt  education-num  \
 0            40.0               1.0  49.0  101320.0           12.0

    capital-gain  capital-loss  hours-per-week  salary
 0           0.0        1902.0            40.0     1.0  ,
 tensor(1),
 tensor([0.2312, 0.7688]))
```

Figure 7.2 – Output of running a prediction on the saved adult_sample_model model

In the web deployment of the model described in this recipe, you will be going through exactly the same steps (as outlined in the following list) as you went through in the recipe from *Chapter 3, Training Models with Tabular Data*, that we just reviewed:

1. Load the saved, trained model.

2. Apply the model to an input sample.

3. Get the prediction from the model.

Unlike the recipe in *Chapter 3, Training Models with Tabular Data*, where all the action took place in the context of a Jupyter notebook, in this recipe you will be accomplishing these steps through a simple web application. You will be able to enter new input samples and get predictions on them in a very natural fashion, and you will see the predictions as clear English statements rather than as tensors. Even better, you will be able to share your web application with others so they can exercise your model and see the predictions that it makes. In short, by deploying your model, you will transform it from an abstract coding artifact that can only be accessed in a program to a useful piece of software that non-programmers can actually use.

The deployment described in this section incorporates a web server implemented as a Flask module. Flask is a Python library that lets you serve web applications from the familiar surroundings of Python. In this recipe, you will start the Flask module and then use the web pages that it serves to exercise the model.

Getting ready

Ensure that you have followed the steps in the *Setting up fastai on your local system* recipe to get fastai installed on your local system. Confirm that you can access the files in the `deploy_tabular` directory of the `ch7` directory of your repo.

How to do it...

To deploy a model trained on a tabular dataset on your system, you will start the Flask server and work through the associated web pages to validate that you can get a prediction from the model for a given set of input scoring parameters. Complete the following steps to do this:

1. From a command window/terminal on your local system, make `deploy_tabular` in the `ch7` directory of your repo your current directory.

2. Enter the following command in the command line/terminal to start the Flask server:

   ```
   python web_flask_deploy.py
   ```

   The output of this command indicates that the web server is running `localhost:5000`, as shown in the following screenshot:

```
PS C:\personal\packt\deploy_test> python web_flask_deploy.py
 * Serving Flask app "web_flask_deploy" (lazy loading)
 * Environment: production
   WARNING: This is a development server. Do not use it in a production deployment.
   Use a production WSGI server instead.
 * Debug mode: on
 * Restarting with windowsapi reloader
 * Debugger is active!
 * Debugger PIN: 217-661-682
 * Running on http://0.0.0.0:5000/ (Press CTRL+C to quit)
```

Figure 7.3 – Output when the Flask server starts

3. Open a browser window and enter the following in the address field:

```
Localhost:5000
```

If the Flask server started successfully, the home.html web page will be loaded in the browser, as shown in *Figure 7.4*:

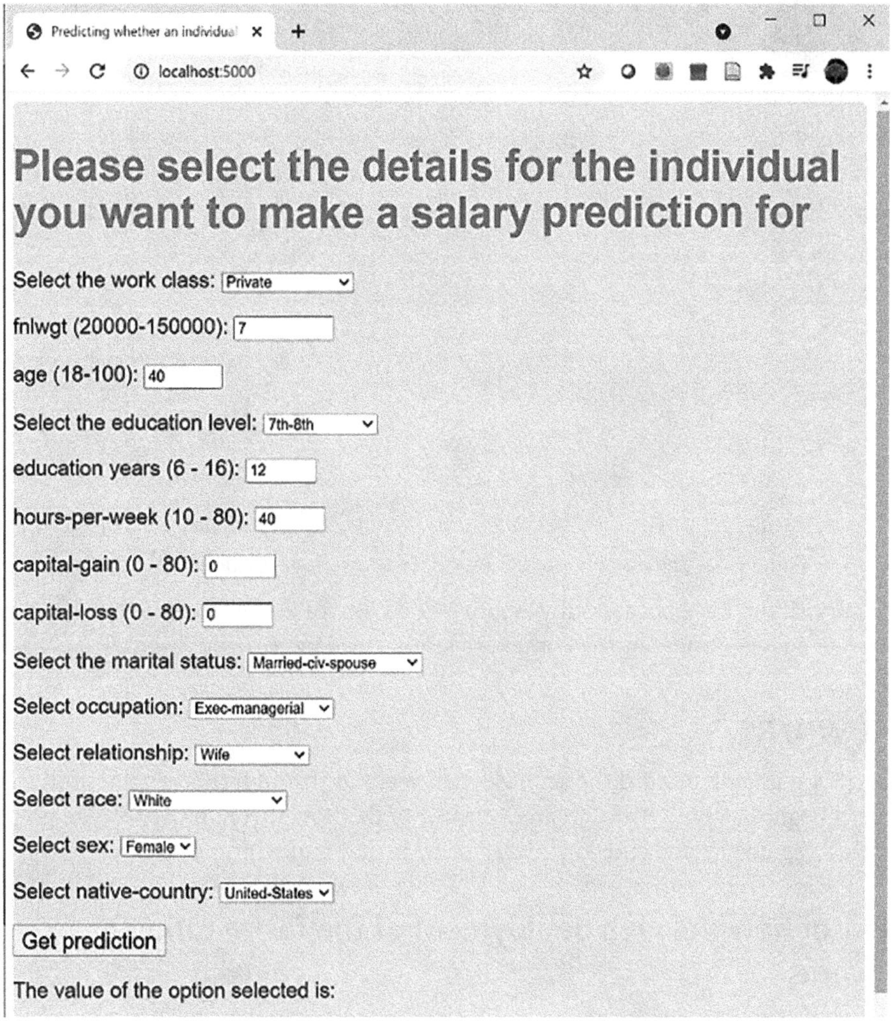

Figure 7.4 – home.html being server by the Flask server

4. Now, select **Get prediction**. When you select this button, the **query string** shown in *Figure 7.5* gets displayed momentarily at the bottom of the page. This query string specifies the values of the **scoring parameters**, that is, the values that will be used to exercise the model. These are the values specified in the fields in home.html, in this case the default values for the fields:

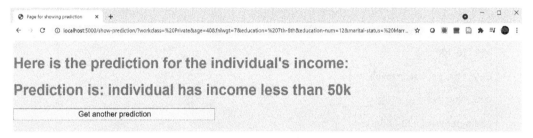

Figure 7.5 – The query string generated with the default setting in home.html

After a few seconds, the show-prediction.html web page is displayed with the prediction the model made for the values entered in home.html, as shown in *Figure 7.6*:

Figure 7.6 – The model's prediction displayed in show-prediction.html

Congratulations! You have successfully set up a Flask server and exercised the web deployment of a fastai model in the context of simple web page deployment.

How it works...

There was a lot going on behind the scenes when you ran through this recipe. In this section, we'll begin by going through an overview of the flow through the web deployment and then we'll dig deeper into the key pieces of code that make up the deployment.

Overview of how the web deployment of the fastai tabular model works

The web deployment described in this recipe is a departure from the recipes you have seen so far in this book. Unlike the other recipes, which involved a single code file in the form of a Jupyter notebook, the web deployment incorporates code that is spread across a series of files, as shown in *Figure 7.7*:

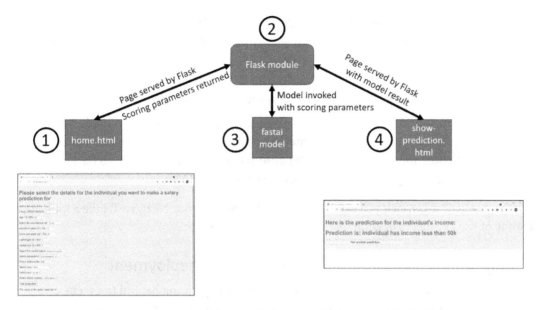

Figure 7.7 – Overview of the web deployment of a fastai model using Flask

Here are the key items highlighted by the numbers in *Figure 7.7*:

1. home.html – This is the web page where the user specifies the **scoring parameters**, that is, the details (including **work class**, **age**, **education level**, and **occupation**) for the individual for whom they want the model to make a prediction. These scoring parameters will be fed into the model and the model will return a prediction on whether the individual described by the scoring parameters has an income above or below 50,000. There is one control on home.html for every feature that was used to train the model. home.html incorporates a set of JavaScript functions that set the available values in each control, package up the user's entries, and call show-prediction.html with the scoring parameters as arguments.

2. The Flask web_flask_deploy.py module – A Python module that uses the Flask library to serve the web pages that make up the web deployment. This module includes **view functions** for home.html and show-prediction. html that do most of the work for the web deployment. The view function for show-prediction.html parses the scoring parameters sent from home. html, assembles the scoring parameter values into a DataFrame, calls the trained model using the DataFrame containing the scoring parameters to get a prediction, generates a string from the model's prediction, and finally triggers show-prediction.html to be displayed with the prediction string.

3.　The fastai `adult_sample_model.pkl` model – This is the model that you trained and saved in the *Saving a trained tabular model* recipe of *Chapter 3, Training Models with Tabular Data*. The view function for `show-prediction.html` in the `web_flask_deploy.py` Flask module loads this model and then uses it to make a prediction with the scoring parameters entered in `home.html`.

4.　`show-prediction.html` – This web page displays the prediction made by the model on the scoring parameters entered in `home.html`. Once the user has read the prediction, they can select the **Get another prediction** button to go back to `home.html` to enter another set of scoring parameters.

That is a high-level summary of how the web deployment works. Next, we'll take a look at some of the key pieces of code that make up the deployment.

Digging deeper into the code behind the web deployment

Now that you have seen the high-level picture of what's happening behind the scenes, let's dig into two pieces of code that are particularly important to the overall web deployment. We'll go through the major code items that make up the deployment, including the Python code in the Flask server module as well as the JavaScript functions in the HTML files.

When you start the Flask server, as shown in *Step 2* of the recipe, the trained model is loaded into the Python module, as shown in this snippet from the Flask server code:

```
path = Path(os.getcwd())
full_path = os.path.join(path,'adult_sample_model.pkl')
learner = load_learner(full_path)
```

Here are the key parts of this snippet:

- `path = Path(os.getcwd())` – Sets `path` to be the directory where you started the Flask server. The code assumes the model file is in the same directory.

- `full_path = os.path.join(path,'adult_sample_model.pkl')` – Defines the full path for the model, including the filename.

- `learner = load_learner(full_path)` – Loads the model into `learner`.

When you go to `localhost:5000` in your browser, the `home.html` page gets displayed. How does this happen? In the `web_flask_deploy.py` Flask module, the **view function** for `home.html` controls what happens when you navigate to `localhost:5000` when the Flask server is active, as shown in the following code snippet:

```
@app.route('/')
def home():
    title_text = "fastai deployment"
    title = {'titlename':title_text}
    return render_template('home.html',title=title)
```

Here are the key parts of this view function:

- `@app.route('/')` – Specifies that this view function is applied when you navigate to the `localhost:5000` address

- `return render_template('home.html',title=title)` – Specifies that `home.html` is displayed when you navigate to `localhost:5000`

As `home.html` is loaded, the action moves from Python in the Flask server module to a combination of HTML and JavaScript in `home.html`. First, the `load_selections()` function is called to load values into the controls in the web page, as shown in the following HTML statement:

```
<body onload="load_selections()">
```

The `load_selections()` function populates the select controls (drop-down lists) on the page with lists specifying the valid values, such as the following for the `relationship` control:

```
var relationship_list = [" Wife" ," Not-in-family" ,"
Unmarried" ," Husband" ," Own-child" ," Other-relative" ];
```

The `load_selections()` function also includes `for` loops that set the values in the select controls to the lists of valid values, such as the following `for` loop that populates the `relationship` control:

```
for(var i = 0; i < relationship_list.length; i++) {
    var opt = relationship_list[i];
    select_relationship.innerHTML += "<option value=\"" + opt
+ "\">" + opt + "</option>";
```

For the controls for entering numeric values, the `load_selections()` function sets the default values that appear when the page is loaded. For example, the following statement in the `load_selections()` function sets the default value for the `age` field:

```
document.getElementById("age").defaultValue = 40;
```

Once the values are loaded in the controls and the page is displayed, the user can select values for the scoring parameters in the controls that are different from the defaults.

After the user has selected values for the scoring parameters, the user can select the **Get prediction** button. Here is the HTML associated with that button. It specifies that the `link_with_args()` function gets called when the button is selected:

```
<button>
<a onclick="link_with_args();" style="font-size : 20px; width:
100%; height: 100px;">Get prediction</a>
</button>
```

The `link_with_args()` function calls the `getOption()` function that loads the values the user selected in the controls in `home.html` and builds the query string with these values, as shown in the following snippet from `getOption()`:

```
prefix = "/show-prediction/?"
window.output = prefix.concat("workclass=",workclass_
string,"&age=",age_value,"&fnlwgt=",fnlwgt_
value,"&education=",education_string,"&education-
num=",education_num_value,"&marital-status=",marital_
status_string,"&occupation=",occupation_
string,"&relationship=",relationship_string,"&race=",race_
string,"&sex=",sex_string,"&capital-gain=",capital_gain_
value,"&capital-loss=",capital_loss_value,"&hours-per-
week=",hours_per_week_value,"&native-country=",native_country_
string);
document.querySelector('.output').textContent = window.output;
```

Here are the key parts of this snippet:

- `prefix = "/show-prediction/?"` – Specifies which view function in the Flask module will be invoked when the link is triggered.

- `window.output` – Specifies the set of parameters included in the query string. This string is made up of a series of key value pairs, where each value equals the corresponding control in `home.html`.

- `document.querySelector('.output').textContent = window.output;` – Specifies that the query string is displayed in the browser window.

You may recall seeing the query string in this recipe. In *Step 4* of the recipe, when you selected the **Get prediction** button in home.html, the query string was briefly displayed at the bottom of the page before show-prediction.html was loaded.

After calling getOption(), the link_with_args() function triggers a reference to show-prediction.html with the following statement:

```
window.location.href = window.output;
```

With this statement, the action switches from the world of HTML and JavaScript back to Python, and the view function for show-prediction.html is invoked in the Flask server. Here is the beginning of this view function where the scoring parameter values that were entered in home.html and passed along in the query string are loaded into the score_df DataFrame:

```
@app.route('/show-prediction/')
def show_prediction():
    score_df = pd.DataFrame(columns=scoring_columns)
for col in scoring_columns:
        print("value for "+col+" is: "+str(request.args.
get(col)))
        score_df.at[0,col] = request.args.get(col)
```

Here are the key parts of this snippet:

- `@app.route('/show-prediction/')` – Specifies that this view function is applied for the show-prediction.html web page.

- `score_df = pd.DataFrame(columns=scoring_columns)` – Creates an empty dataframe to hold the scoring parameters.

- `score_df.at[0,col] = request.args.get(col)` – This statement is run for every column in the scoring_columns list. It copies the values in the query string that was built in the getOption() JavaScript function and passed as part of the reference to show-prediction.html to the corresponding column in the first row of the score_df dataframe. This is how the scoring parameter values that the user entered in home.html are brought into the Python Flask server module.

Now that the scoring parameters have been loaded into the first row of the `score_df` dataframe, we can invoke the model on the first row of the dataframe, as shown in the following code snippet from the view function for `show-prediction.html`:

```
pred_class,pred_idx,outputs = learner.predict(score_df.iloc[0])
if outputs[0] >= outputs[1]:
        predict_string = "Prediction is: individual has income
less than 50k"
    else:
        predict_string = "Prediction is: individual has income
greater than 50k"
    prediction = {'prediction_key':predict_string}
    return(render_template('show-prediction.
html',prediction=prediction))
```

Here are the key parts of this snippet:

- `pred_class,pred_idx,outputs = learner.predict(score_df.iloc[0])` – Invokes the model with the first row of the `score_df` DataFrame as input. The call has three outputs:

  a) `pred_class` lists the scoring parameters as they were fed into the model. For the categorical columns, the original scoring parameter value is replaced with the category identifier. For example, the value `United States` in the `native-country` column is replaced with `40.0`. These transformations are exactly the same as the transformations that were done on the training data back when you trained the model in *Chapter 3, Training Models with Tabular Data*. Thanks to the way that fastai manages these transformations, unlike Keras, you don't need to worry about maintaining a pipeline object and applying it when you deploy a model – fastai just takes care of it. This is a great advantage of fastai.

  b) `pred_idx` – The index of the prediction. For this model, the prediction will be either 0 (indicating the individual has an income less than 50,000) or 1 (indicating the individual has an income more than 50,000).

  c) `outputs` – Shows the probability of each prediction value.

  Figure 7.8 shows an example of prediction output and how it corresponds to the `pred_class`, `pred_idx`, and `outputs` variables:

```
(   workclass  education  marital-status  occupation  relationship  race  sex  \
0        5.0       8.0             3.0         0.0           6.0   5.0  1.0

    native-country  education-num_na   age     fnlwgt  education-num  \          ◄━━━ pred_class
0             40.0              1.0  49.0   101320.0           12.0

    capital-gain  capital-loss  hours-per-week  salary
0           0.0        1902.0            40.0     1.0  ,
 tensor(1),                                                                      ◄━━━ pred_idx
 tensor([0.2312, 0.7688]))                                                       ◄━━━ outputs
```

Figure 7.8 – Example of model prediction output

- `return(render_template('show-prediction. html',prediction=prediction))` – Specifies that `show-prediction. html` is displayed with the argument value set in this view function.

With this statement, the action moves back to HTML as `show-prediction.html` is loaded in the browser. The following snippet shows the HTML that displays the prediction text:

```html
<div class="home">
  <h1 style="color: green">
    Here is the prediction for the individual's income:
  </h1>
  <h1 style="color: green">
    {{ prediction.prediction_key }}
  </h1>
```

The `{{ prediction.prediction_key }}` value corresponds to the `predict_ string` value that was set in the view function for `show-prediction` in the Flask server. The result is that the prediction that the model made on the scoring parameters is displayed, as shown in *Figure 7.9*:

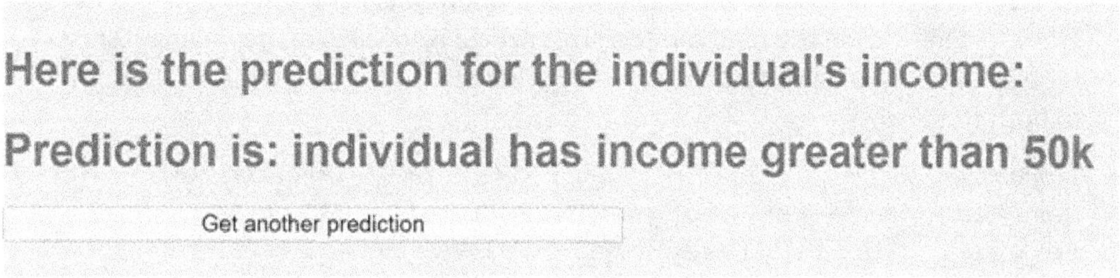

Here is the prediction for the individual's income:

Prediction is: individual has income greater than 50k

Get another prediction

Figure 7.9 – The end result of the deployed model – a prediction on the scoring parameters

Now you have seen all the major code items that make up the entire flow of this web deployment of a fastai model. The flow goes through the following steps:

1. The flow begins when you start the Flask server. Once you have started the Flask server, it is ready to serve `home.html` at `localhost:5000`.

2. When you go to `localhost:5000` in your browser, the view function for `home.html` runs in the Flask server and `home.html` is displayed in the browser.

3. The flow then goes to HTML/JavaScript in `home.html`, where the user selects the scoring parameters and selects the **Get prediction** button.

4. The flow then returns to the Flask server where the view function for `show-prediction.html` is run to get a prediction from the model on the scoring parameters and to display `show-prediction.html` in the browser.

5. Finally, the flow returns back to HTML in `show-prediction.html`, where the model's prediction is displayed.

6. At this point, the user can select the **Get another prediction** button in `show-prediction.html` to start the process over again at *Step 2* with a different set of scoring parameters.

There's more...

The tour through the web deployment example in this recipe only scratches the surface of what you can do with Flask, and it covers only the bare minimum of what you can do with modern HTML and JavaScript. A thorough exploration of how to develop web applications with Python is beyond the scope of this book, but if you are interested in learning more, you can check out the following resources:

*   *Deploying a Deep Learning Model using Flask* (`https://towardsdatascience.com/deploying-a-deep-learning-model-using-flask-3ec166ef59fb`) goes into additional details about how to use Flask to deploy deep learning models. This article is focused on deploying Keras models rather than fastai models, but the principles described in the article apply to both frameworks.

*   *Responsive Web Design with HTML5 and CSS* (`https://www.amazon.com/Responsive-Web-Design-HTML5-CSS/dp/1839211563/ref=sr_1_2?dchild=1&keywords=html5+packt&qid=1623055650&sr=8-2`) gives a broad background on modern HTML and also covers **cascading style sheets** (**CSS**), which are used to control how web pages are rendered.

- *Clean Code in JavaScript* (`https://www.amazon.com/Clean-Code-JavaScript-reliable-maintainable/dp/1789957648/ref=sr_1_6?dchild=1&keywords=Javascript+Packt&qid=1623055616&sr=8-6`) describes good programming practices for JavaScript. If you are reading this book, you are reasonably comfortable with Python, and you should not have much trouble picking up JavaScript. JavaScript doesn't always get the respect it deserves from people who program in more illustrious languages such as C++ and Scala, but the fact is that JavaScript is remarkably flexible and really useful to know.

Deploying a fastai model trained on an image dataset

In the *Deploying a fastai model trained on a tabular dataset* recipe, we went through the process of deploying a model trained on a tabular dataset. We deployed a model that predicted whether an individual would have an income over 50,000 based on a set of characteristics called **scoring parameters**, including education, job category, and hours worked per week. To do this deployment, we needed a way to allow the user to select values for the scoring parameters and then show the prediction made by the trained fastai model on these scoring parameters.

In this recipe, we will deploy the image classification model that you trained in the *Training a classification model with a standalone vision dataset* recipe of *Chapter 6, Training Models with Visual Data*. This model predicts what fruit or vegetable is depicted in an image. Unlike the deployment of the tabular dataset model, to deploy the image dataset model we will need to be able to specify an image file on which to make a prediction.

> **Note**
> For the sake of simplicity, this deployment uses web pages with the same names (`home.html` and `show-prediction.html`) as the deployment of a tabular dataset that we did in the *Deploying a fastai model trained on a tabular dataset* recipe. However, these web pages are customized for the image model deployment.

Getting ready

Ensure that you have followed the steps in the *Setting up fastai on your local system* recipe to get fastai installed on your local system. Confirm that you can access the files in the `deploy_image` directory of the `ch7` directory of your repo.

How to do it...

To exercise the deployment of an image classification model on your local system, you will start the Flask server, open up the home.html page for this deployment in your browser, select an image file to get a prediction on, and then validate that you get a prediction for the image displayed in the show-prediction.html page for this deployment.

Complete the following steps to exercise the deployment of a fastai model trained on an image dataset:

1. From a command window/terminal on your local system, make deploy_image in the ch7 directory of your repo your current directory.

2. Enter the following command in the command line/terminal to start the Flask server:

    ```
    python web_flask_deploy_image_model.py
    ```

    This output of this command indicates that the web server is running localhost:5000, as shown in *Figure 7.10*:

```
PS C:\personal\packt\deploy_image> python web_flask_deploy_image_model.py
path is: C:\personal\packt\deploy_image
full_path is:  C:\personal\packt\deploy_image\fruits_360may3.pkl
 * Serving Flask app "web_flask_deploy_image_model" (lazy loading)
 * Environment: production
   WARNING: This is a development server. Do not use it in a production deployment.
   Use a production WSGI server instead.
 * Debug mode: on
 * Restarting with windowsapi reloader
path is: C:\personal\packt\deploy_image
full_path is:  C:\personal\packt\deploy_image\fruits_360may3.pkl
 * Debugger is active!
 * Debugger PIN: 217-661-682
 * Running on http://0.0.0.0:5000/ (Press CTRL+C to quit)
```

Figure 7.10 – Output when the Flask server starts

3. Open a browser window and enter the following in the address field:

    ```
    localhost:5000
    ```

    If the Flask server started successfully, the web page home.html will be loaded in the browser, as shown in *Figure 7.11*:

Figure 7.11 – home.html for the image model deployment being served by the Flask server

4. Now select **Choose File**. A file dialog opens. In this dialog, navigate to the `test_` `images` subdirectory in your `deploy_images` directory. Select the lemon image file, `5_100.jpg`, and close the file dialog, for example, by selecting **Open** in Windows.

5. When the file dialog closes, the name of the file you selected shows in `home.html` beside the **Choose File** button, as shown in *Figure 7.12*:

Figure 7.12 – The name of the file you selected displayed in home.html

6. Now select **Get prediction**. The `show-prediction.html` page is displayed with the model's prediction for what is displayed in the image you selected in `home.html`, as shown in *Figure 7.13*:

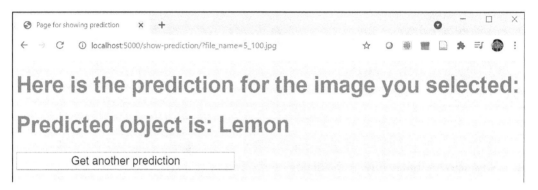

Figure 7.13 – The model's prediction of what is in the image, shown in show-prediction.html

Congratulations! You have successfully set up a Flask server and exercised the web deployment of a fastai model that predicts the object depicted in an image.

How it works...

Now that you have worked through the web deployment of a fastai image classification model, let's go through what's happening behind the scenes. We'll start by going through an overview of the deployment and then digging into the code differences between the deployment of the image classification model and the deployment of a tabular dataset model as described in the *Deploying a fastai model trained on a tabular dataset* recipe.

Overview of how the web deployment of the fastai image classification model works

Let's review the end-to-end flow of the deployment, as shown in *Figure 7.14*:

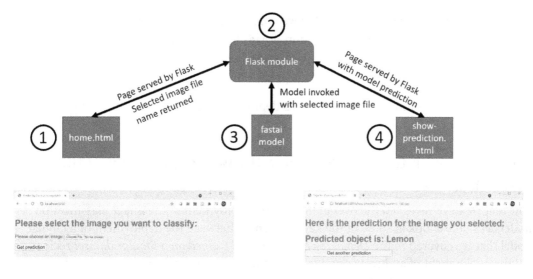

Figure 7.14 – Overview of web deployment of a fastai image classification model using Flask

Here are the key items highlighted by the numbers in *Figure 7.14*:

1. home.html – This is the web page where the user specifies the image file that they want the model to make a prediction on. The version of home.html used for the image classification model deployment incorporates HTML and JavaScript functions that display a file selection dialog, package up the name of the selected file, and call show-prediction.html with the name of the selected image file as an argument.

2. The Flask web_flask_deploy_image_model.py module – The Python module that uses the Flask library to serve the web pages that make up the web deployment. This module includes view functions for home.html and show-prediction.html. The view function for show-prediction.html ingests the name of the image file selected in home.html, calls the trained model using this filename to get a prediction, generates a string from the model's prediction, and finally triggers show-prediction.html to be displayed with the prediction string.

3. The fastai fruits_360may3.pkl image classification model – This is the model that you trained and saved in the *Training a classification model with a standalone vision dataset* recipe of *Chapter 6, Training Models with Visual Data*. The view function for show-prediction.html in the web_flask_deploy_image_model.py Flask module loads this model and then uses it to make a prediction for the image file selected in home.html.

4. `show-prediction.html` – This web page displays the prediction made by the model on the image file selected in `home.html`. On this page, the user can optionally select the **Get another prediction** button to go back to `home.html` to select another image file to make a prediction on.

That is a high-level summary of how the web deployment of the image classification model works.

Digging deeper into the code behind the web deployment of the image classification model

Now that we have reviewed the overall flow of the deployment of the image classification model, let's take a look at some of the key differences between the deployment of a tabular model that we covered in the *Deploying a fastai model trained on a tabular dataset* recipe and the deployment of an image classification model that we worked through in this section. Here are the key differences:

- HTML in `home.html` – The version of `home.html` for the deployment of the tabular dataset model required a large number of controls to allow the user to specify all the required scoring parameters. The user needed to be able to specify values for all the features that were used to train the model. For the deployment of the image classification model, there is only one feature – the image file – so we only need a control for selecting a file. Here is the HTML for the file selection control:

```
<label for="image_field">Please choose an image:</
label>
    <input type="file"
        id="image_field" name="image_field"
        accept="image/png, image/jpeg">
```

Here are the key items in this piece of HTML:

a) `input type="file"` – Specifies that this control is for inputting files from the local filesystem.

b) `accept="image/png, image/jpeg"` – Specifies that the file dialog that is opened from this control will only allow image files with `image/png` or `image/jpeg` content types to be selected.

- JavaScript in `home.html` – The version of `home.html` for the deployment of the tabular dataset model has three JavaScript functions:

a) `getOption()`, to get the values from the controls

b) `link_with_args()`, to call `getOption()` and send the query string to the view function for `show-prediction.html`

c) `load_selections()`, to initialize the controls

The version of `home.html` used for the image classification model deployment doesn't need `load_selections()` (because there aren't any controls that need to initialized) and its version of `link_with_args()` is identical to the tabular model deployment version. That leaves the `getOption()` function, which is significantly different from the version in the tabular model deployment. Here is the image classification deployment version of `getOption()`:

```
function getOption() {
    var file_value = [];
    const input = document.querySelector('input');
    const curFiles = input.files;
    if(curFiles.length === 0) {
        console.log("file list empty");

    } else {
    for(const file of curFiles) {
        file_value.push(file.name);
      }
    }
    prefix = "/show-prediction/?"
    window.output = prefix.concat("file_name=",file_
value[0])
    }
```

Here are the key items in the definition of `getOption()`:

a) `const input = document.querySelector('input');` – Associates `input` with the file selector.

b) `const curFiles = input.files;` – Assigns the list of files associated with the file selector to `curFiles`.

c) `for (const file of curFiles) { file_value.push(file.name);}` – Loops through the files in the file list associated with the file selector and adds each filename to the `file_value` list.

d) `window.output = prefix.concat("file_name=",file_value[0])` – Builds the query string using the first element from the `file_value` list of filenames. We will only make predictions on one file at a time, so we just need one filename for the query string. The resulting query string would look something like this: `/show-prediction/?file_name=5_100.jpg`.

- View function for `show-prediction.html` in the Flask server – The following code snippet shows what this view function looks like for the image classification web deployment:

```
@app.route('/show-prediction/')
def show_prediction():
    image_file_name = request.args.get("file_name")
    full_path = os.path.join(path,image_directory,image_file_name)
    img = PILImage.create(full_path)
    pred_class, ti1, ti2 = learner.predict(img)
    predict_string = "Predicted object is: "+pred_class
    prediction = {'prediction_key':predict_string}
    return(render_template('show-prediction.html',prediction=prediction))
```

Here are the key items in the definition of this view function:

a) `image_file_name = request.args.get("file_name")` – Sets the value of `image_file_name` to be the filename from the query string.

b) `full_path = os.path.join(path,image_directory,image_file_name)` – Sets the value of `full_path` to be the fully qualified filename for the image file selected in `home.html`. This assumes that the file was selected from the `test_images` subdirectory of the directory where you are running the Flask server.

c) `img = PILImage.create(full_path)` – Creates an image object called `img` for the image file selected in `home.html`.

d) `pred_class, ti1, ti2 = learner.predict(img)` – Gets a prediction from the image classification model for the `img` object. `pred_class` contains the category (such as *Apple* or *Pear*) predicted by the model for the image file.

e) `return(render_template('show-prediction.
html',prediction=prediction))` – Specifies that `show-prediction.
html` is displayed with the argument value set in this view function.

Now you have seen all the major code differences between the deployment of the tabular dataset model and the deployment of the image classification model.

There's more...

In this chapter, you have seen two examples of deploying fastai models using a Flask-based web application. This isn't the only approach that you can take to deploy models. Other approaches include deploying models via REST API endpoints (so that other applications can invoke the models directly) or encapsulating models along with their dependencies in other applications. Models can be packaged with the dependencies (such as required Python libraries) in Docker containers, and then these containers can be made available to other applications through orchestration systems such as Kubernetes.

Rather than dwelling on these general deployment concepts, it may be more useful to review some concrete approaches to deploying fastai models in particular. Here is a sample of approaches to deploying fastai models:

- Deploying with Amazon SageMaker, the AWS machine learning environment, as described here: `https://aws.amazon.com/blogs/machine-learning/building-training-and-deploying-fastai-models-with-amazon-sagemaker/`. This approach requires using some PyTorch code directly and may not have been validated on the latest version of fastai.

- Deploying in AWS using TorchServe, as described here: `https://aws.amazon.com/blogs/opensource/deploy-fast-ai-trained-pytorch-model-in-torchserve-and-host-in-amazon-sagemaker-inference-endpoint/`. This approach has fewer *moving parts* than the approach described in the previous point, and it is more current, but the essence of the approach seems to be re-implementing the fastai model in PyTorch.

- Deploying using Google Cloud Platform, as described here: `https://jianjye.medium.com/how-to-deploy-fast-ai-models-to-google-cloud-functions-for-predictions-e3d73d71546b`.

- Deploying using Azure, as described here: `https://forums.fast.ai/t/platform-azure/65527/7`.

This list is not exhaustive but it does demonstrate the variety of deployment options available for fastai models.

Maintaining your fastai model

Deploying a model is not the end of the story. Once you have deployed a model, you need to maintain the deployment so that it matches the current characteristics of the data on which the model is trained. A thorough description of how to maintain a deep learning model in production is beyond the scope of this book, but it is worthwhile to touch on how to maintain models in the context of the simple model deployments described in this chapter. In this recipe, we will look at actions you could take to maintain the tabular model that you deployed in the *Deploying a fastai model trained on a tabular dataset* recipe.

Getting ready

Ensure that you have followed the steps in the *Setting up fastai on your local system* recipe to get fastai installed on your local system. Also ensure that you have the Flask server started for the tabular model deployment by following *Steps 1, 2, and 3* from the *Deploying a fastai model trained on a tabular dataset* recipe.

In this recipe, you will be doing some basic analysis of the training data that you used to train the tabular model deployed in the *Deploying a fastai model trained on a tabular dataset* recipe. To prepare for this analysis, confirm that you can use your spreadsheet of choice (such as Excel or Google Sheets) to open up adult.csv, the file in the ADULT_ SAMPLE dataset that contains the training data. If you don't already have adult.csv on your local system, follow these steps to get this file on your local system and confirm you can open it up with your spreadsheet application:

1. In your Gradient environment, enter the following command in a terminal window to copy adult.csv to your temp directory:

   ```
   cp /storage/data/adult_sample/adult.csv /notebooks/temp/
   adult.csv
   ```

2. In JupyterLab in your Gradient environment, navigate to the temp directory where you copied adult.csv in the previous step, right-click on adult.csv, and select **Download**.

3. Use your spreadsheet application to open up the local copy of adult.csv that you downloaded in the previous step. *Figure 7.15* shows what the first few rows of adult.csv look like in Excel:

	A	B	C	D	E	F	G	H	I	J	K	L	M	N	O
1	age	workclass	fnlwgt	education	education-	marital-sta	occupatio	relationshi	race	sex	capital-gai	capital-los	hours-per-	native-cou	salary
2	49	Private	101320	Assoc-acd	12	Married-civ-spouse		Wife	White	Female	0	1902	40	United-St	>=50k
3	44	Private	236746	Masters	14	Divorced	Exec-man	Not-in-fa	White	Male	10520	0	45	United-St	>=50k
4	38	Private	96185	HS-grad		Divorced			Unmarrie	Black	Female	0	0	32	United-St	<50k
5	38	Self-emp	112847	Prof-scho	15	Married-c	Prof-speci	Husband	Asian-Pac-	Male	0	0	40	United-St	>=50k
6	42	Self-emp	82297	7th-8th		Married-c	Other-ser	Wife	Black	Female	0	0	50	United-St	<50k
7	20	Private	63210	HS-grad	9	Never-ma	Handlers-	Own-child	White	Male	0	0	15	United-St	<50k
8	49	Private	44434	Some-coll	10	Divorced		Other-rel	White	Male	0	0	35	United-St	<50k
9	37	Private	138940	11th	7	Married-civ-spouse		Husband	White	Male	0	0	40	United-St	<50k
10	46	Private	328216	HS-grad	9	Married-c	Craft-repa	Husband	White	Male	0	0	40	United-St	>=50k
11	36	Self-emp	216711	HS-grad		Married-civ-spouse		Husband	White	Male	99999	0	50	?	>=50k
12	23	Private	529223	Bachelors	13	Never-married		Own-child	Black	Male	0	0	10	United-St	<50k
13	18	Private	216284	11th		Never-ma	Adm-cleri	Own-child	White	Female	0	0	20	United-St	<50k
14	30	Private	151989	Assoc-voc		Married-civ-spouse		Wife	White	Female	0	0	40	United-St	<50k

Figure 7.15 – The first few rows of adult.csv in Excel

> **Note**
>
> You may wonder why I am suggesting using a spreadsheet to examine the data in this recipe. Why not use Python? There are a couple of reasons I recommend a spreadsheet here. First, no less an authority than Jeremy Howard stated that Excel is a great data science tool, and I happen to think he is absolutely right. It's flexible, lightweight, and faster than Python for the kind of simple investigation on a small dataset that I'm featuring in this recipe. Second, Excel helped me to debug a problem with the deployment of the tabular model. When I first tested the deployment, I struggled to understand why the deployed model produced different predictions than the model invoked in a Python notebook. However, once I examined the data in Excel, the problem was obvious: all categorical values began with spaces in the data used for training the model. The categorical values that users could select in the deployment didn't start with spaces, so the model did not recognize them as being the same as the categorical values it had encountered at training time. Excel gave me a quick way to detect the root cause of the problem.

How to do it...

To exercise some model maintenance actions, complete the following steps:

1. First, take a closer look at how categorical values are represented in ADULT_
   SAMPLE. If you don't already have your local copy of adult.csv open in your
   spreadsheet application, open it up now. Select one of the values in the workclass
   column. Do you notice anything unusual about the value? Check out the values in
   some of the other categorical columns: relationship and native-country.
   You will see that the values in every categorical column begin with a blank.

2. Recall that in home.html, the user is constrained in the choices they can make for each of the categorical features. Open up home.html in the tabular model deployment and see what values are available for workclass. *Figure 7.16* shows the values that the user can select for workclass:

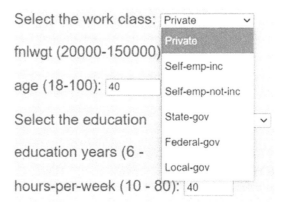

Figure 7.16 – Values available for workclass in home.html

3. The values that users can choose in home.html for the categorical columns are defined in a series of lists in the load_selections() JavaScript function. Here are the lists defined in the load_selections() JavaScript function for workclass, relationship, and native-country:

```
var workclass_list = [" Private" ," Self-emp-inc" ,"
Self-emp-not-inc" ," State-gov" ," Federal-gov" ," Local-
gov" ];
var relationship_list = [" Wife" ," Not-in-family" ,"
Unmarried" ," Husband" ," Own-child" ," Other-relative"
];
var native_country_list = [" United-States"," Puerto-
Rico"," Mexico"," Canada"," Taiwan"," Vietnam","
Philippines"];
```

Notice how the values in each of these lists begin with a blank, just like the values in the corresponding categorical columns in adult.csv. The values in these lists are used to build the query string, which in turn gets used as input to get a prediction from the model in the view function for show-prediction.html. What would have happened if the values in the lists in home.html had been defined without leading blanks?

4. Suppose that the training data for the model gets expanded to include individuals who come from the **United Kingdom**, and you received an updated version of adult.csv that includes rows with the value United-Kingdom in the native-country column. Here is what you would need to do to update the deployment to adapt to this change:

a) Retrain the model with the new version of adult.csv and save the new trained model to a pkl file using the learner.export() fastai API. For the purposes of this recipe, suppose that you call the new adult_sample_model_new.pkl model.

b) Copy the updated adult_sample_model_new.pkl model file into the deploy_tabular directory on your local system.

c) Update the definition of the model path in the web_flask_deploy.py Flask server module to include the new model filename:

```
full_path = os.path.join(path, 'adult_sample_model_new.
pkl')
```

d) Update native_country_list in home.html to include the new value:

```
var native_country_list = [" United-States"," Puerto-
Rico"," Mexico"," Canada"," Taiwan"," Vietnam","
Philippines", "  United-Kingdom"  ];
```

You would need to take the same steps to update the deployment for new values in any of the categorical columns: retrain the model with the updated training dataset, copy the updated trained model to the deployment directory, update the Flask server so that it loads the updated model, and update the list of valid categorical values in home.html.

5. In the previous step, we saw what we will need to do if the dataset gets expanded with new categorical values. What if a brand-new column is added to the dataset? Just like the process described in *Step 4*, you would need to retrain the model on the updated training dataset that includes the new column, copy the new model into the deploy_tabular directory, and update web_flask_deploy.py so that it loads the new model. Finally, you would need to update home.html to allow the user to enter information about the new column. The updates you need to make depend on whether the new column is **continuous**, that is, contains numeric values, or **categorical**, that is, contains a fixed set of values. The following two steps describe the specific updates you would need to make to home.html in each case.

6. Suppose you need to update the deployment to handle a new continuous column
    called `years-in-job` – a count of how many years the individual has been in
    their current job. The valid values are 0 to 45, and the default value is 5. To add this
    column, you need to make several updates to `home.html`. First, you need to add
    the following code to define a control for this new column:

    ```html
    <p>
        <label for="years-in-job">years in job (0 - 45):</
    label>
        <input type="number" id="years-in-job" name="years-in-
    job"  min="0" max="45">
        </p>
    ```

    Next, you need to add the following line to the `load_selection()` JavaScript
    function to set the default value:

    ```javascript
    document.getElementById("years-in-job").defaultValue = 5;
    ```

    Next, you need to add the following line to the `getOption()` JavaScript function
    to set the value that will be included in the query string for this column:

    ```javascript
    years_in_job_value = document.getElementById("years-in-
    job ").value;
    ```

    Finally, you need to add the following to the end of the query string used to define
    `window.output`:

    ```javascript
    ,"&years-in-job=",years_in_job_value
    ```

7. Suppose you need to update the deployment to handle a new categorical column
    called `work-location` that specifies the location of the individual's current job.
    Valid values for this column are `remote`, `on-site`, and `mixed`. To adapt `home.
    html` to work with this new column, start by adding the following code to define
    a control for the `work-location` column:

    ```html
    <p>
        Select work location:
        <select id="work-location">
        </select>
    </p>
    ```

Next, add the following lines to the `load_selection()` JavaScript function to set the values for the control for the new column. Note that we assume that like the other categorical columns, values in `work-location` will be preceded by a blank, so the values in `work_location_list` all begin with blanks:

```
var select_work_location = document.getElementById("work-
location");

var work_location_list = [" remote"," on-site"," mixed"];

for(var i = 0; i < work_location_list.length; i++) {
    var opt = work_location_list[i];
    select_work_location.innerHTML += "<option
value=\"" + opt + "\">" + opt + "</option>";
    }
```

Next, add the following lines to the `getOption()` JavaScript function to set the value that will be included in the query string for this column:

```
selectElementworklocation = \
document.querySelector('#work-location');
work_location_string =\
selectElementworklocation.
options[selectElementworklocation.selectedIndex].value
```

Finally, add the following to the end of the query string used to define `window.output`:

```
,"&work-location=",work_location_string
```

Congratulations! You have worked through some of the actions required to maintain the deployment of a model to ensure it can adapt to changes in the training dataset.

How it works...

In this recipe, we reviewed how you could maintain the web deployment of a model trained on tabular data. We saw the steps we would take to adapt the deployment to work with changes in the training dataset. The dataset changes we covered were new values in existing categorical columns, new continuous columns, and new categorical columns.

In an industrial-strength production deployment, the schema of the dataset, that is, the characteristics of all the columns that make up the dataset, would be maintained outside of the HTML files. For example, we might maintain the schema in a separate configuration file that had information about the columns in the dataset. Instead of being hardcoded, the controls and valid values in `home.html` would be built on the fly using the contents of the configuration file.

With this kind of dynamic setup, when a new column is added to the dataset or the valid values for a column change, we will only have to update the definition of the dataset schema in the configuration file and `home.html` would be updated automatically. To keep the web deployment as easy to follow as possible, we code the controls and valid values directly in `home.html` instead of building them dynamically. This made the *Deploying a fastai model trained on a tabular dataset* recipe easier to follow, but it also meant there were several places in `home.html` that had to be updated to maintain the deployment when the dataset changed.

There's more...

In this recipe, we talked about how to handle changes to the dataset schema, but we didn't talk about how to deal with changes in the distribution of the dataset or how to monitor the model to ensure that it maintained its performance over time. Both of these issues are critical to maintaining a deployed model but they are beyond the scope of this book.

If you are interested in learning more about monitoring the performance of models in production, this article provides a great overview: `https://christophergs.com/machine%20learning/2020/03/14/how-to-monitor-machine-learning-models/`.

Test your knowledge

Now that you have deployed two kinds of fastai models and worked through some of the challenges related to maintaining deployed models, you can try some additional variations on deployment to exercise what you've learned.

Getting ready

Ensure that you have followed the steps in the *Setting up fastai on your local system* recipe to get fastai installed on your local system. Also, ensure that you have the Flask server started for the image classification model deployment by following *Steps 1, 2,* and *3* in the *Deploying a fastai model trained on an image dataset* recipe.

To experiment on the image classification model deployment, make a copy of the `deploy_image` directory. To do this, make the directory that contains `deploy_image` your current directory and run the following command to make a copy of the directory and its contents called `deploy_image_test`:

```
cp -r deploy_image deploy_image_test
```

How to do it...

You can follow the steps in this recipe to extend and enhance the model deployment that you followed in the *Deploying a fastai model trained on an image dataset* recipe to allow the user to select multiple image files in `home.html` and show predictions for all the images in `show-prediction.html`:

1. Make `deploy_image_test` your current directory.

2. To allow users to select multiple files and show predictions for all the files at once, you will need to update `home.html`, the Flask server, and `show-prediction.html`.

3. Start by updating `home.html` so that the user can select multiple files in the file dialog. Add the `multiple` attribute to the definition of the file dialog control, as shown in the following HTML snippet:

```
<input type="file" multiple
       id="image_field" name="image_field"
       accept="image/png, image/jpeg">
```

Now the user will be able to select more than one file in the file dialog.

4. Next, update the `getOption()` JavaScript function in `home.html` to build a list of filenames to add to the query string and send back to the Flask server. The updated `getOption()` function looks like this:

```
function getOption() {
    var file_value = [];
    var file_count = 0;
    const input = document.querySelector('input');
    var file_path = input.value;
    const curFiles = input.files;
    if(curFiles.length === 0) {
        console.log("file list empty");
    } else {
```

```
    for(const file of curFiles) {
        if (file_count == 0) {
            file_count = 1;
            file_list_prefix = "&file_name=";
            var file_list = file_list_prefix.concat(file.name);
        } else {
            file_list = file_list.concat("&file_name=",file.
    name);
        }
        file_value.push(file.name);
    }
    }
    prefix = "/show-prediction/?"
    window.output = prefix.concat("file_path=",file_
    path,file_list)
    }
```

Here are the key updated items in the getOption() function:

a) `var file_list = file_list_prefix.concat(file.name);` –
   Specifies that if this is the first file, initialize the `file_list` string

b) `file_list = file_list.concat("&file_name=",file.name);`
   – Specifies that if this isn't the first file, add the filename to the end of the `file_`
   `list` string

c) `window.output = prefix.concat("file_path=",file_`
   `path,file_list)` – Specifies that the query string includes the `file_list`
   string, which has the filenames for all the image files selected by the user

You have completed the updates required in home.html to handle multiple
image files.

5. Now it's time to update the Flask server. First, add the following function to the
    Flask server. You will use this function later to build the parameters that you will
    send to show-prediction.html:

```
def package_list(key_name,list_in):
    i = 0
    list_out = []
    for element in list_in:
        key_value = list_in[i].strip()
```

```
            list_out.append({key_name:key_value})
            i = i+1
        return(list_out)
```

6. Next, update the view function for show-prediction.html. First, you will
    want to bring the list of filenames that you built in the getOption() function of
    home.html into a Python list. The following statement will create such a list called
    image_file_name_list:

    ```
    image_file_name_list = request.args.getlist('file_name')
    ```

7. Next, update the view function for show-prediction.html so that you iterate
    through image_file_name_list to get a prediction for each file in the list.
    Save the pred_class value for each prediction in a list called prediction_
    string_list.

8. Use the package_list function that you defined in *Step 5* to prepare
    prediction_string_list to send to show-prediction.html:

    ```
        prediction_list = package_list("prediction_
    key",prediction_string_list)
    ```

9. Update the return statement of the view function to include
    prediction_list:

    ```
        return(render_template('show-prediction.
    html',prediction_list=prediction_list))
    ```

    Now you have completed the updates to the Flask server required to handle multiple
    image files.

10. Next, update show-prediction.html to show the predictions for each of
    the images:

    ```
        <h1 style="color: green">
            Here are the predictions for the images you selected:
        </h1>
        <h1 style="color: green">
        <p>
        {% for prediction in prediction_list %}
            {{prediction.prediction_key}}{% if not loop.last %},
    {% endif %}
        {% endfor %}
    ```

```
      </p>
      </h1>
```

11. Now test whether everything works. Start the Flask server in `deploy_image_test`:

```
python web_flask_deploy_image_model.py
```

12. Go to `localhost:5000` in your browser to display `home.html`. Select the **Choose Files** button to bring up the file selection dialog. In the file selection dialog, select the `4_100.jpg`, `5_100.jpg`, and `26_100.jpg` files from the `deploy_image_test/test_images` directory. Once you have selected these files, `home.html` will be updated to indicate that three files have been selected, as shown in *Figure 7.17*:

Please select the images you want to classify:

Get prediction

Figure 7.17 – home.html after selecting three image files

13. Select the **Get prediction** button. You should now see the predictions for all three files in `show-predictions.html`, as shown in *Figure 7.18*:

Here are the predictions for the images you selected:

Apricot, Lemon, Beetroot

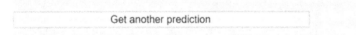

Figure 7.18 – show-prediciton.html showing predictions for multiple images

Congratulations! You have completed a useful extension to the deployment of the image classification model.

8
Extended fastai and Deployment Features

So far in this book, you have learned how to ingest and explore datasets using fastai, how to train fastai models with tabular, text, and image datasets, and how to deploy fastai models. Throughout the book so far, the emphasis has been on covering as much of the functionality of fastai as possible using the highest-level fastai API. In particular, we have emphasized using `dataloaders` objects as the basis for defining the datasets used to train the model. Up to this point in the book, we have taken the *happy path* whenever possible. To demonstrate how to accomplish tasks using fastai, we have chosen the most straightforward way possible.

In this chapter, we are going to take some steps off the *happy path* to explore additional features of fastai. You will learn how to track what is happening with your model more closely, how to control the training process, and generally how to take advantage of more of the capabilities that fastai has to offer. We are also going to cover some more advanced topics related to model deployment.

Here are the recipes that will be covered in this chapter:

- Getting more details about models trained with tabular data
- Getting more details about image classification models
- Training a model with augmented data
- Using callbacks to get the most out of your training cycle
- Making your model deployments available to others
- Displaying thumbnails in your image classification model deployment
- Test your knowledge

Technical requirements

In this chapter, you will be using both your cloud environment and your local environment for model deployment:

- Ensure that you have completed the setup sections from *Chapter 1, Getting Started with fastai*, and have a working Gradient instance or Colab setup.
- Ensure that you have completed the steps described in the *Setting up fastai on your local system* recipe in *Chapter 7, Deployment and Model Maintenance*, to set up fastai on your local system.

Ensure that you have cloned the repo for the book from `https://github.com/PacktPublishing/Deep-Learning-with-fastai-Cookbook` and have access to the `ch8` folder. This folder contains the code samples described in this chapter.

Getting more details about models trained with tabular data

In the *Training a model in fastai with a curated tabular dataset* recipe of *Chapter 3, Training Models with Tabular Data*, you trained a fastai model on a tabular dataset and used accuracy as the metric. In this recipe, you will learn how to get additional metrics for this model: **precision** and **recall**. Precision is the ratio of true positives divided by true positives plus false positives. Recall is the ratio of true positives divided by true positives plus false negatives.

These are useful metrics. For example, the model we are training in this recipe is predicting whether an individual's income is over 50,000. If it is critical to avoid false positives – that is, predicting an income over 50,000 when the individual has an income less than that – then we want precision to be as high as possible. This recipe will show you how to add these useful metrics to the training process for a fastai model.

Getting ready

Confirm that you can open the `training_with_tabular_datasets_metrics.ipynb` notebook in the `ch8` directory of your repo.

How to do it...

In this recipe, you will be running through the `training_with_tabular_datasets_metrics.ipynb` notebook. Once you have the notebook open in your fastai environment, complete the following steps:

1. Run the cells in the notebook up to the `Define and train model` cell to import the required libraries, set up your notebook, and prepare the dataset.

2. Run the following cell to define and train the model:

```
recall_instance = Recall()
precision_instance = Precision()
learn = tabular_learner(dls,layers=[200,100],
metrics=[accuracy,recall_instance,precision_instance])
learn.fit_one_cycle(3)
```

Here are the key items in this cell:

a) `recall_instance = Recall()` – defines a recall metric object. Note that you will get an error if you put `Recall` directly in the metrics list for the model. Instead, you need to define a recall metric object, such as `recall_instance`, and include that object in the metrics list. See the fastai documentation (`https://docs.fast.ai/metrics.html#Recall`) for more details on this metric.

b) `precision_instance = Precision()` – defines a precision metric object. You will get an error if you put `Precision` directly in the metrics list, so you need to define the `precision_instance` object first and then include that object in the metrics list for the model.

c) `metrics=[accuracy,recall_instance,precision_instance]` – specifies that the model will be trained with accuracy, recall, and precision as metrics.

The output of this cell, as shown in *Figure 8.1*, includes accuracy as well as recall and precision for each epoch of the training run:

epoch	train_loss	valid_loss	accuracy	recall_score	precision_score	time
0	0.337531	0.354399	0.830508	0.609375	0.722222	00:07
1	0.325458	0.356652	0.826271	0.562500	0.734694	00:07
2	0.317219	0.344203	0.838983	0.578125	0.770833	00:07

```
CPU times: user 21.2 s, sys: 112 ms, total: 21.3 s
Wall time: 21.3 s
```

Figure 8.1 – Training output including recall and precision

Congratulations! You have trained a model with tabular data and generated recall and precision metrics for the training process.

How it works...

You may have asked yourself how I knew that you could not include `Recall` and `Precision` directly in the metrics list for the model and that you needed to define objects first and then include those objects in the metrics list. The simple answer is that it was trial and error. More specifically, when I attempted to include `Recall` and `Precision` directly in the metrics list, I got the following error:

```
TypeError: unsupported operand type(s) for *: 'AccumMetric' and 'int'
```

When I searched for this error, I came across this post in the fastai forum: https://forums.fast.ai/t/problem-with-f1scoremulti-metric/63721. The post explained the reason for the error and that to get around it I needed to define the `Recall` and `Precision` objects first and then include them in the metrics list.

This experience is an example of both a weakness and a strength of fastai. The weakness is that the documentation for `Precision` and `Recall` is missing an essential detail – you cannot use them directly in the metrics list. The strength is that the fastai forum provides clear and accurate resolutions for issues like this and demonstrates the strength of the fastai community.

Getting more details about image classification models

In the *Training a classification model with a simple curated vision dataset* recipe of *Chapter 6, Training Models with Visual Data*, you trained an image classification model using the CIFAR curated dataset. The code to train and exercise the model was straightforward because we took advantage of the highest-level structures in fastai. In this recipe, we will revisit this image classification model and explore techniques in fastai to get additional information about the model and its performance, including the following:

- Examining the **pipeline** that fastai generates to prepare the data
- Getting a chart of the training and validation loss during the training process
- Displaying the images where the model performs worst
- Displaying the **confusion matrix** to get a snapshot of where the model is not doing well
- Applying the model to the test set and examining the model's performance on the test set

In this recipe, we are going to expand the recipe where we trained the CIFAR curated dataset. By taking advantage of the additional features of fastai, we will be able to understand our model better.

Getting ready

Confirm that you can open the training_with_image_datasets_datablock. ipynb notebook in the ch8 directory of your repo.

How to do it...

In this section, you will be running through the training_with_image_datasets_ datablock.ipynb notebook. Once you have the notebook open in your fastai environment, complete the following steps:

1. Update the following cell to ensure that model_path points to a writeable directory in your Gradient or Colab instance:

   ```
   model_path = '/notebooks/temp'
   ```

2. Run the cells in the notebook up to the Define a DataBlock cell to import the required libraries, set up your notebook, and ingest the CIFAR dataset.

3. Run the following cell to define a `DataBlock` object. By defining a `DataBlock` object explicitly, we will be able to do additional actions that we couldn't do directly on a `dataloaders` object, such as getting a summary of the pipeline:

```
db = DataBlock(blocks = (ImageBlock, CategoryBlock),
                get_items=get_image_files,
                splitter=RandomSplitter(seed=42),
                get_y=parent_label)
```

Here are the key items in this cell:

a) `blocks = (ImageBlock, CategoryBlock)` – specifies that the input to the model is images (`ImageBlock`) and the target is a categorization of the input images (`CategoryBlock`).

b) `get_items=get_image_files` – specifies that the `get_image_files` function is called to get the input to the `DataBlock` object.

c) `splitter=RandomSplitter(seed=42)` – specifies how the validation set is defined from the training set. By default, 20% of the training set is randomly selected to make up the validation set. By specifying a value for `seed`, this call to `RandomSplitter` produces consistent results across multiple runs. See the documentation for `RandomSplitter` (`https://docs.fast.ai/data. transforms.html#RandomSplitter`) for more details.

d) `get_y=parent_label` – specifies that the labels for the images (that is, the categories to which the images belong) are defined by the directories where the images are located in the input dataset. For example, on Gradient, cat images in the training set are found in the `/storage/data/cifar10/ train/cat` directory.

4. Run the following cell to define a `dataloaders` object using the `DataBlock` object `db` that you created in the previous cell:

```
dls = db.dataloaders(path/'train',bs=32)
```

Here are the key items in this cell:

a) `db.dataloaders` – specifies that the `dataloaders` object is created using the `DataBlock` object `db`

b) `path/'train'` – specifies that the input to this model is only the training subset of the `CIFAR` dataset.

5. Run the following cell to get a summary of the pipeline:

```
db.summary(path/"train")
```

Let's look at the key parts of the output of this cell. First, the output shows details about the input dataset, including the source directory, the size of the whole dataset, and the size of the training and validation sets, as shown in the following screenshot:

```
Setting-up type transforms pipelines
Collecting items from /storage/data/cifar10/train
Found 50000 items
2 datasets of sizes 40000,10000
```

Figure 8.2 – Summary description of the input dataset

Next, the output shows the pipeline that fastai applies to a single input sample, including the source directory of the sample, the image object that is created for the sample, and the label (category) for the sample, as shown in *Figure 8.3*:

```
Building one sample
  Pipeline: PILBase.create
    starting from
      /storage/data/cifar10/train/truck/3702_truck.png
    applying PILBase.create gives
      PILImage mode=RGB size=32x32
  Pipeline: parent_label -> Categorize -- {'vocab': None, 'sort': True, 'add_na': False}
    starting from
      /storage/data/cifar10/train/truck/3702_truck.png
    applying parent_label gives
      truck
    applying Categorize -- {'vocab': None, 'sort': True, 'add_na': False} gives
      TensorCategory(9)
```

Figure 8.3 – Summary description of the pipeline for one image file

Next, the output shows the pipeline that fastai applies to build a single batch, that is, converting the image objects that are output from the sample pipeline into tensors. As shown in *Figure 8.4*, the 32 x 32-pixel image objects are converted to 3 x 32 x 32 tensors, where the first dimension contains color information about the image:

```
Building one batch
Applying item_tfms to the first sample:
  Pipeline: ToTensor
    starting from
      (PILImage mode=RGB size=32x32, TensorCategory(9))
    applying ToTensor gives
      (TensorImage of size 3x32x32, TensorCategory(9))
```

Figure 8.4 – Summary description of the pipeline applied to a single batch

Finally, the output shows the transformations applied to the batches as a whole, as shown in *Figure 8.5*:

```
Applying batch_tfms to the batch built
  Pipeline: IntToFloatTensor -- {'div': 255.0, 'div_mask': 1}
    starting from
      (TensorImage of size 4x3x32x32, TensorCategory([9, 7, 9, 0], device='cuda:0'))
    applying IntToFloatTensor -- {'div': 255.0, 'div_mask': 1} gives
      (TensorImage of size 4x3x32x32, TensorCategory([9, 7, 9, 0], device='cuda:0'))
```

Figure 8.5 – Summary description of the pipeline applied to all batches

6. Run the following cell to define a `DataBlock` object for the test set:

```
db_test = DataBlock(blocks = (ImageBlock, CategoryBlock),
                    get_items=get_image_files,
                    splitter=RandomSplitter(valid_
pct=0.99,seed=42),
                    get_y=parent_label)
```

Note that unlike the `DataBlock` object for the training set, we define `db_test` with an explicit value for `valid_pct`. We set this value to 99% because we won't be doing any training of the model when we apply the test set to it, so there is no need to hold back any of the test set for training. We don't set `valid_pct` to `1.0` because that value will produce an error when you apply a summary to `db_test`.

7. Run the cells in the notebook up to the `Define and train the model` cell to examine the dataset.

8. Run the following cell to define the model with a `cnn_learner` object. Note that because you defined a `dataloaders` object from the `DataBlock` object you get the best of both worlds: the additional features (such as a summary) available only with a `DataBlock` object along with the familiar code pattern for `dataloaders` objects that you have used for most of the recipes in this book:

```
learn = cnn_learner(dls, resnet18,
                    loss_
func=LabelSmoothingCrossEntropy(),
                    metrics=accuracy)
```

9. Run the following cell to train the model:

```
learn.fine_tune(2,cbs=ShowGraphCallback())
```

Note the `cbs=ShowGraphCallback()` parameter. With this parameter, the output of the training process includes a graph of training and validation loss, as shown in *Figure 8.6*:

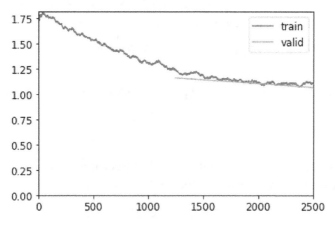

Figure 8.6 – Training and validation loss graph

This graph contains the same data as the table of training results that you get by default from the training process, as shown in *Figure 8.7*:

epoch	train_loss	valid_loss	accuracy	time
0	1.248233	1.160732	0.714300	01:50
1	1.109203	1.067518	0.752600	01:49

Figure 8.7 – Training and validation loss table

10. Run the following cell to save the trained model. We update the path for the model temporarily to a directory that is writeable in Gradient so that we can save the model:

```
save_path = learn.path
learn.path = Path(model_path)
learn.save('cifar_save_'+modifier)
learn.path = save_path
```

Here are the key items in this cell:

a) `save_path = learn.path` – specifies that the current path for the model is saved to `save_path`.

b) `learn.path = Path(model_path)` – specifies that the path for the model is set to a writeable directory.

c) `learn.save('cifar_save_'+modifier)` – saves the model. We will load the saved model later to exercise the model with the test set.

d) `learn.path = save_path` – resets the path for the model to its original value.

11. Run the following cell to confirm the performance of the trained model in terms of accuracy:

```
learn.validate()
```

The second value in the output should match the accuracy you saw in the final epoch of the training, as shown in *Figure 8.8*:

```
(#2) [1.0675183534622192,0.7526000142097473]
```

Figure 8.8 – Output of validate

12. Run the cells up to the `Examine the top loss examples and confusion matrix` cell.

13. Run the following cell to see the samples where the model has the biggest loss:

```
interp = ClassificationInterpretation.from_learner(learn)
interp.plot_top_losses(9, figsize=(15,11))
```

Here are the key items in this cell:

a) `interp = ClassificationInterpretation.from_learner(learn)` – specifies that `interp` is an interpretation object for the `learn` model

b) `interp.plot_top_losses(9, figsize=(15,11))` – specifies that the nine images with the highest losses should be displayed

The output shows examples of the images where the model has the biggest loss along with the predicted contents of the image and the actual contents of the image. You can think of these as the images where the model did the worst job predicting what's in the images. *Figure 8.9* shows a subset of the output. For example, for the first displayed image, the model predicted the image contained a bird while the image is actually labeled as a cat:

Prediction/Actual/Loss/Probability

bird/cat / 6.00 / 0.97   cat/deer / 4.98 / 0.96   automobile/ship / 4.85 / 0.95

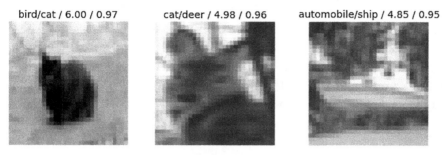

Figure 8.9 – Examples of images with the most loss

14. Run the following cell to generate a confusion matrix for the trained model:

```
interp.plot_confusion_matrix()
```

The output of this cell is a confusion matrix that summarizes the performance of the trained model, as shown in *Figure 8.10*. A confusion matrix is an *N* x *N* matrix, where *N* is the number of target classes. It compares the actual target class values (the vertical axis) with the predicted values (the horizontal axis). The diagonal of the matrix shows the cases where the model made the correct prediction, while all the entries off the diagonal are cases where the model made an incorrect prediction. For example, in the confusion matrix shown in *Figure 8.10*, in 138 instances the model predicted that an image of a dog was a cat, and in 166 instances it predicted that an image of a cat was a dog:

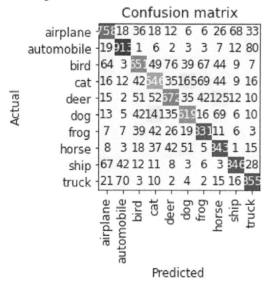

Figure 8.10 – Confusion matrix for the trained model

15. Now that you have examined the performance of the model with the training set, let's examine how the model does with the test set. To do this, we will define a new `dataloaders` object using the test set, define a model with this `dataloaders` object, load the saved weights from the trained model, and then do the same steps to evaluate the model performance that we did with the model trained on the training set. To begin, run the following cell to create a new `dataloaders` object `dls_test` that is defined with the test dataset:

```
dls_test = db_test.dataloaders(path/'test',bs=32)
```

16. Run the following cell to define a new model object, `learn_test`, that is based on the `dataloaders` object you created in the previous step. Note that the model definition is identical to the model you defined for the training set in *Step 8* except that it uses the `dataloaders` object `dls_test` that was defined with the test dataset in the previous step:

```
learn_test = cnn_learner(dls_test, resnet18,
                     loss_
func=LabelSmoothingCrossEntropy(),
                     metrics=accuracy)
```

17. Run the following cell to load the saved weights from the model trained with the training set:

```
learn_test.path = Path(model_path)
learn_test.load('cifar_save_'+modifier)
```

Here are the key items in this cell:

a) `learn_test.path = Path(model_path)` – specifies that the path for the `learn_test` model is changed to the directory where the model weights were saved in *Step 10*

b) `learn_test.load('cifar_save_'+modifier')` – specifies that the `learn_test` model gets loaded with the weights from the model trained with the training set

Now we are all set to exercise the model with the test set.

18. Run the following cell to see the overall accuracy of the model on the test set:

```
learn_test.validate()
```

The second value in the output is the accuracy of the model on the test set, as shown in *Figure 8.11*:

```
(#2) [1.0432653427124023,0.7565000057220459]
```

Figure 8.11 – Output of validate on the test set

19. Run the following cell to see the images in the test set where the model has the biggest loss:

```
interp_test = ClassificationInterpretation.from_
learner(learn_test)
interp_test.plot_top_losses(9, figsize=(15,11))
```

The output shows examples of the images in the test set where there was the biggest loss along with the predicted contents of the image and the actual contents of the image in the test set. *Figure 8.12* shows a subset of the output:

Prediction/Actual/Loss/Probability

cat/bird / 4.54 / 0.71   dog/bird / 4.44 / 0.83   cat/frog / 4.37 / 0.68

  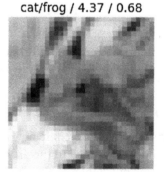

Figure 8.12 – Sample images from the test set where the model performed worst

20. Run the following cell to get the confusion matrix for the model applied to the test set:

```
interp_test.plot_confusion_matrix()
```

The output of this cell is a confusion matrix that summarizes the performance of the model on the test set, as shown in *Figure 8.13*. Note that the numbers in this confusion matrix are smaller than the numbers in the confusion matrix for the model applied to the training set:

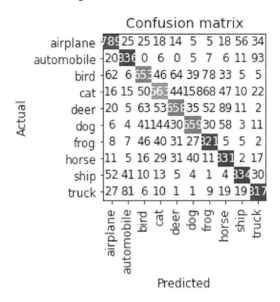

Figure 8.13 – Confusion matrix for the model applied to the test set

Congratulations! You have worked through an image classification model to see the benefit of additional information that fastai can provide. You have also learned how to apply the model to the entire test set and examine the model's performance on the test set.

How it works...

It's instructive to compare the model we created in the recipe in this section with the model we created in the *Training a classification model with a simple curated vision dataset* recipe of *Chapter 6, Training Models with Visual Data*.

Here is the definition of the `dataloaders` object from the *Chapter 6, Training Models with Visual Data*, recipe:

```
dls = ImageDataLoaders.from_folder(path, train='train',
valid='test')
```

And here is the definition of the `dataloaders` object from this recipe. Unlike the previous `dataloaders` definition, this definition makes use of the `DataBlock` object db:

```
dls = db.dataloaders(path/'train',bs=32)
```

Here is the definition of the `DataBlock` object db:

```
db = DataBlock(blocks = (ImageBlock, CategoryBlock),
               get_items=get_image_files,
               splitter=RandomSplitter(seed=42),
               get_y=parent_label)
```

What is the benefit of defining the `dataloaders` object using the `DataBlock` object?

First, by starting with a `DataBlock` object, you have more control over the details of how the dataset is set up. You can explicitly define the function that defines the input dataset (the function assigned to `get_items`) as well as the function that defines the label (the function assigned to `get_y`). You may recall that we took advantage of this flexibility in the *Training a multi-image classification model with a curated vision dataset* recipe of *Chapter 6, Training Models with Visual Data*. In that recipe, we needed to ensure that the input dataset excluded images that had no annotation. By using a `DataBlock` object in that recipe, we were able to define a custom function to assign to `get_items` that excluded images with no annotation.

Second, we can take advantage of some additional functions in fastai if we have a `DataBlock` object. In this recipe, we were able to apply the `summary()` function to the `DataBlock` object to see the pipeline that fastai applied to the input dataset. The `summary()` function cannot be applied to a `dataloaders` object, so we would have missed out on the additional details about the data pipeline if we had not defined a `DataBlock` object.

If a `DataBlock` object is so useful, why didn't we use one in the *Training a classification model with a simple curated vision dataset* recipe of *Chapter 6, Training Models with Visual Data*? We only used a `dataloaders` object in that recipe (instead of starting with a `DataBlock` object) because that recipe was relatively simple – we didn't need the additional flexibility of a `DataBlock` object. Throughout this book, we have stuck with the highest-level APIs for fastai whenever we could, including in that recipe. Simplicity is a key benefit of fastai, so if it's possible to go with the highest-level APIs (including using `dataloaders` directly), it makes sense to keep it simple and stick with the highest-level APIs.

Training a model with augmented data

In the previous recipe, you learned about some additional facilities that fastai provides to keep track of your model and you learned how to apply the test set to the model trained on the training set. In this recipe, you will learn how to combine these techniques with another technique that fastai makes it easy to incorporate in your model training: **data augmentation**. With data augmentation, you can expand your training dataset to include variations on the original training samples and, potentially, improve the performance of the trained model.

Figure 8.14 shows some results of augmentation applied to an image from the CIFAR dataset:

Figure 8.14 – Augmentation applied to an image

You can see in these examples that the augmentations applied to the image include flipping the image on the vertical axis, rotating the image, and adjusting the brightness of the image.

As in the previous recipe, in this recipe, we are going to work with an image classification model trained on the CIFAR curated dataset, but in this recipe, we will experiment with augmenting the images in the dataset.

Getting ready

Confirm that you can open the training_with_image_datasets_datablock_ augmented.ipynb notebook in the ch8 directory of your repo.

How to do it...

In this recipe, you will be running through the `training_with_image_datasets_` `datablock_augmented.ipynb` notebook. Here is a high-level summary of what you will do in this recipe:

1. Train the model with non-augmented data.
2. Train the model with augmented data.
3. Exercise the model trained with non-augmented data with the test set.
4. Exercise the model trained with augmented data with the test set.

Once you have the notebook open in your fastai environment, complete the following steps:

1. Update the following cell to ensure that `model_path` points to a writeable directory in your Gradient or Colab instance:

   ```
   model_path = '/notebooks/temp'
   ```

2. Run the cells in the notebook up to the `Try augmenting the training set` cell to import the required libraries, set up your notebook, and train a model on the non-augmented `CIFAR` dataset.

3. Run the following cell to create a new `DataBlock` object db2 that incorporates augmentation and a `dataloaders` object dls2 that is defined with this new `DataBlock` object:

   ```
   db2 = db.new(batch_tfms=aug_transforms())
   dls2 = db2.dataloaders(path/'train',bs=32)
   ```

   Here are the key items in this cell:

   a) `db2 = db.new(batch_tfms=aug_transforms())` – defines a new `DataBlock` object db2 that incorporates the default augmentation defined by `aug_transforms()`. See the fastai documentation for details on `aug_transforms()`: `https://docs.fast.ai/vision.augment.` `html#aug_transforms`.

   b) `dls2 = db2.dataloaders(path/'train',bs=32)` – defines a new `dataloaders` object dls2 based on the `DataBlock` object db2. Now, dls2 is defined with the training subset of the input dataset.

4. Run the following cell to get a summary of the pipeline:

```
db2.summary(path/"train")
```

The output of this cell gives us details about the pipeline that is applied to the dataset. First, the output shows details about the input dataset, including the source directory, the size of the whole dataset, and the size of the training and validation sets, as shown in *Figure 8.15*:

```
Setting-up type transforms pipelines
Collecting items from /storage/data/cifar10/train
Found 50000 items
2 datasets of sizes 40000,10000
```

Figure 8.15 – Summary description of the input dataset

Next, the output shows the pipeline that fastai applies to a single input sample, including the source directory of the sample, the image object that is created for the sample, and the label (category) for the sample, as shown in *Figure 8.16*:

```
Building one sample
  Pipeline: PILBase.create
    starting from
      /storage/data/cifar10/train/truck/3702_truck.png
    applying PILBase.create gives
      PILImage mode=RGB size=32x32
  Pipeline: parent_label -> Categorize -- {'vocab': None, 'sort': True, 'add_na': False}
    starting from
      /storage/data/cifar10/train/truck/3702_truck.png
    applying parent_label gives
      truck
    applying Categorize -- {'vocab': None, 'sort': True, 'add_na': False} gives
      TensorCategory(9)
```

Figure 8.16 – Summary description of the pipeline for one image file

Next, the output shows the pipeline that fastai applies to build a single batch, that is, converting the image objects that are output from the sample pipeline into tensors. As shown in *Figure 8.17*, the 32 x 32-pixel image objects are converted to 3 x 32 x 32 tensors, where the first dimension contains color information about the image:

```
Building one batch
Applying item_tfms to the first sample:
  Pipeline: ToTensor
    starting from
      (PILImage mode=RGB size=32x32, TensorCategory(9))
    applying ToTensor gives
      (TensorImage of size 3x32x32, TensorCategory(9))
```

Figure 8.17 – Summary description of the pipeline applied to a single batch

These first three portions of the output are identical to the same portions of the output of summary() from the *Getting more details about image classification models* recipe. The final portion, however, is different and describes the transformations that are applied to images in the augmentation process, as shown in *Figure 8.18*:

```
Applying batch_tfms to the batch built
  Pipeline: IntToFloatTensor -- {'div': 255.0, 'div_mask': 1} -> Flip -- {'size': None, 'mode': 'bilinear', 'pad_mode': 'refl
ection', 'mode_mask': 'nearest', 'align_corners': True, 'p': 0.5} -> Brightness -- {'max_lighting': 0.2, 'p': 1.0, 'draw': No
ne, 'batch': False}
    starting from
      (TensorImage of size 4x3x32x32, TensorCategory([9, 7, 9, 0], device='cuda:0'))
    applying IntToFloatTensor -- {'div': 255.0, 'div_mask': 1} gives
      (TensorImage of size 4x3x32x32, TensorCategory([9, 7, 9, 0], device='cuda:0'))
    applying Flip -- {'size': None, 'mode': 'bilinear', 'pad_mode': 'reflection', 'mode_mask': 'nearest', 'align_corners': Tr
ue, 'p': 0.5} gives
      (TensorImage of size 4x3x32x32, TensorCategory([9, 7, 9, 0], device='cuda:0'))
    applying Brightness -- {'max_lighting': 0.2, 'p': 1.0, 'draw': None, 'batch': False} gives
      (TensorImage of size 4x3x32x32, TensorCategory([9, 7, 9, 0], device='cuda:0'))
```

Figure 8.18 – Description of the pipeline (including augmentation transformations) applied to all batches

5. Run the following cell to see examples of the augmentation transformations being applied to an image in the dataset:

```
dls2.train.show_batch(unique=True, max_n=8, nrows=2)
```

The unique=True argument specifies that we want to see examples of augmentations applied to a single image.

An example of the output of this cell is shown in *Figure 8.19*: note the variations in the augmented images, including being flipped on the vertical axis, having varying brightness, and having varying degrees of vertical space taken up by the truck:

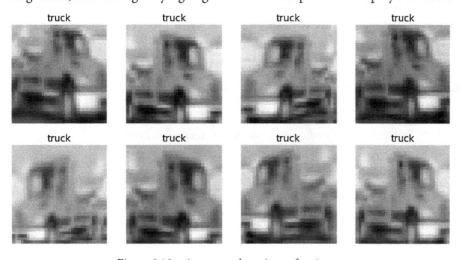

Figure 8.19 – Augmented versions of an image

6. Run the following cell to define a model to be trained with the augmented dataset:

```
learn2 = cnn_learner(dls2, resnet18,
                    loss_
func=LabelSmoothingCrossEntropy(),
                    metrics=accuracy)
```

7. Run the following cell to train the model with the augmented dataset:

```
learn2.fine_tune(2)
```

The output of this cell shows the training loss, validation loss, and validation accuracy, as shown in *Figure 8.20*:

epoch	train_loss	valid_loss	accuracy	time
0	1.356869	1.199771	0.695700	02:26
1	1.197363	1.099920	0.736000	02:29

Figure 8.20 – Output of training the model with the augmented dataset

8. Run the following cell to save the model that was trained on the augmented dataset:

```
save_path = learn2.path
learn2.path = Path(model_path)
learn2.save('cifar_augmented_save_'+modifier)
learn2.path = save_path
```

9. Run the cells up to the `Examine the performance of the model trained on the augmented dataset on the test set` cell to get an idea of the performance of the model trained on the non-augmented dataset making predictions on the test dataset.

10. Run the following cell to define a `dataloaders` object `dls_test` associated with the test set:

```
dls_test = db_test.dataloaders(path/'test',bs=32)
```

11. Run the following cell to define the `learn_augment_test` model to be exercised on the test dataset:

```
learn_augment_test = cnn_learner(dls_test, resnet18,
                    loss_
func=LabelSmoothingCrossEntropy(),
                    metrics=accuracy)
```

12. Run the following cell to load the `learn_augment_test` model with the weights saved from training the model with the augmented dataset:

```
learn_augment_test.path = Path(model_path)
learn_augment_test.load('cifar_augmented_save_'+modifier)
```

Now we have a `learn_augment_test` learner object that has the weights from training with the augmented dataset and is associated with the test dataset.

13. Run the following cell to get the overall accuracy of the `learn_augment_test` model exercised on the test set:

```
learn_augment_test.validate()
```

The output of this cell shows the accuracy of the model on the test set, as shown in *Figure 8.21*:

```
(#2) [1.1021363735198975,0.7366666793823242]
```

Figure 8.21 – Accuracy of the model trained on the augmented dataset applied to the test set

14. Run the following cell to get examples of the images from the test dataset where the model trained on the augmented dataset has the biggest loss:

```
interp_augment_test = ClassificationInterpretation.from_
learner(learn_augment_test)
interp_augment_test.plot_top_losses(9, figsize=(15,11))
```

The output of this cell shows the images from the test set where the model trained on the augmented dataset has the biggest loss. *Figure 8.22* shows examples from this output:

Figure 8.22 – Examples of images where the model trained on
the augmented dataset has the biggest losses

15. Run the following cell to see the confusion matrix for the model trained on the augmented dataset applied to the test set:

```
interp_augment_test.plot_confusion_matrix()
```

The output of this cell is the confusion matrix shown in *Figure 8.23*:

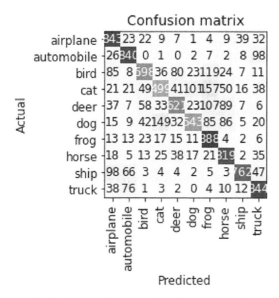

Figure 8.23 – Confusion matrix for the model trained on the augmented dataset applied to the test set

Congratulations! You have trained a fastai image classification model using an augmented dataset and exercised the trained model on the test set.

How it works...

In this recipe, we went through the process of training an image classification model on an augmented dataset. The notebook we worked through in this recipe also included training an image classification on a non-augmented dataset. Let's compare the performance between the models to see whether it was worthwhile to use the augmented dataset.

If we compare the training results between the two models, as shown in *Figure 8.24*, the performance of the model trained on the non-augmented dataset seems to be better:

Training results with non-augmented dataset

epoch	train_loss	valid_loss	accuracy	time
0	1.248233	1.160732	0.714300	02:07
1	1.109203	1.067518	0.752600	02:05

Training results with augmented dataset

epoch	train_loss	valid_loss	accuracy	time
0	1.356869	1.199771	0.695700	02:26
1	1.197363	1.099920	0.736000	02:29

Figure 8.24 – Comparison of training results when training with and without augmented data

Now let's compare the performance of the two models on the test dataset. *Figure 8.25* shows the output of `validate()` for the model trained on the non-augmented and augmented datasets, applied to the test set. Again, the model trained on the non-augmented dataset seems to be better:

Validating on test set (non-augmented training dataset)   Validating on test set (augmented training dataset)

(#2) [1.0601791143417358,0.7543434500694275]   (#2) [1.1021363735198975,0.7366666793823242]

Figure 8.25 – Comparison of validate() results when training with and without augmented data

Figure 8.26 shows the confusion matrixes for the model trained on the non-augmented and augmented datasets, applied to the test set:

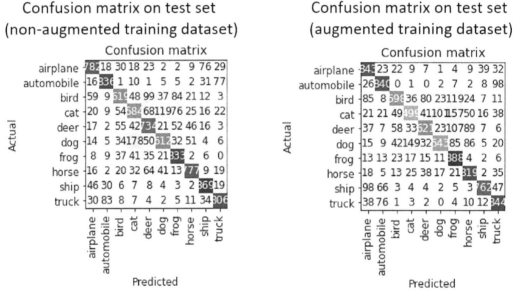

Figure 8.26 – Comparison of confusion matrix when training with and without augmented data

Overall, the model trained with the augmented dataset does not seem to have better performance on the test set than the model trained with the non-augmented data. This may be disappointing, but it's OK.

The purpose of this recipe was to demonstrate to you how to take advantage of the augmented data facilities in fastai, not to do a thorough analysis to get the best possible results with augmented data. Not every application will benefit from augmentation, and we only tried the default augmentation approach. fastai offers a wide range of augmentation options for image models, as described in the fastai documentation: `https://docs.fast.ai/vision.augment.html`. It is possible that this particular dataset had characteristics (such as relatively low resolution) that made augmentation less effective. It's also possible that the set of augmentations included in the default fastai augmentation approach were not ideal for this dataset, and a custom set of augmentations could have produced better results.

How does `aug_transforms()`, the function applied in this recipe, augment the images in the training set? We can get a clue by comparing the pipeline summary for the model trained with the non-augmented training set with the model trained with augmented data. *Figure 8.27* shows the final section of the output of `summary()` for the non-augmented `DataBlock` object:

```
Applying batch_tfms to the batch built
  Pipeline: IntToFloatTensor -- {'div': 255.0, 'div_mask': 1}
    starting from
      (TensorImage of size 4x3x32x32, TensorCategory([9, 7, 9, 0], device='cuda:0'))
    applying IntToFloatTensor -- {'div': 255.0, 'div_mask': 1} gives
      (TensorImage of size 4x3x32x32, TensorCategory([9, 7, 9, 0], device='cuda:0'))
```

Figure 8.27 – Output of summary for the non-augmented DataBlock object

Figure 8.28 shows the final section of the output of `summary()` for the augmented `DataBlock` object:

```
Applying batch_tfms to the batch built
  Pipeline: IntToFloatTensor -- {'div': 255.0, 'div_mask': 1} -> Flip -- {'size': None, 'mode': 'bilinear', 'pad_mode':
'reflection', 'mode_mask': 'nearest', 'align_corners': True, 'p': 0.5} -> Brightness -- {'max_lighting': 0.2, 'p': 1.0,
'draw': None, 'batch': False}
    starting from
      (TensorImage of size 4x3x32x32, TensorCategory([9, 7, 9, 0], device='cuda:0'))
    applying IntToFloatTensor -- {'div': 255.0, 'div_mask': 1} gives
      (TensorImage of size 4x3x32x32, TensorCategory([9, 7, 9, 0], device='cuda:0'))
    applying Flip -- {'size': None, 'mode': 'bilinear', 'pad_mode': 'reflection', 'mode_mask': 'nearest', 'align_corner
s': True, 'p': 0.5} gives
      (TensorImage of size 4x3x32x32, TensorCategory([9, 7, 9, 0], device='cuda:0'))
    applying Brightness -- {'max_lighting': 0.2, 'p': 1.0, 'draw': None, 'batch': False} gives
      (TensorImage of size 4x3x32x32, TensorCategory([9, 7, 9, 0], device='cuda:0'))
```

Figure 8.28 – Output of summary for the non-augmented DataBlock object

By comparing the section of the summary output for these two models, you can understand what transformations are being applied on the augmented dataset, as shown in *Figure 8.29*:

```
Pipeline: IntToFloatTensor -- {'div': 255.0, 'div_mask': 1} -> Flip -- {'size': None, 'mode': 'bilinear', 'pad_mode':
'reflection', 'mode_mask': 'nearest', 'align_corners': True, 'p': 0.5} -> Brightness -- {'max_lighting': 0.2, 'p': 1.0,
'draw': None, 'batch': False}
```

Figure 8.29 – Transformations applied in aug_transforms()

You have now examined some of the differences between training an image classification model with an augmented dataset and training with a non-augmented dataset.

Using callbacks to get the most out of your training cycle

So far in this book, we have kicked off the training process for a fastai model by applying fit_one_cycle or fine_tune to the learner object and have then just let the training run through the specified number of epochs. For many of the models we have trained in this book, this approach has been good enough, particularly for models where each epoch takes a long time and we only train for one or two epochs. But what about models where we want to train the model for 10 or more epochs? If we simply let the training process go to the end, we face the problem shown in the training results shown in *Figure 8.30*. In *Figure 8.30*, we see the result of training a tabular model for 10 epochs with metric set to accuracy:

epoch	train_loss	valid_loss	accuracy	time
0	0.341347	0.360610	0.822034	00:10
1	0.334492	0.369722	0.809322	00:10
2	0.331509	0.331108	0.838983	00:10
3	0.312824	0.337519	0.838983	00:10
4	0.312502	0.347008	0.843220	00:10
5	0.303788	0.319890	0.851695	00:10
6	0.300724	0.315468	0.851695	00:10
7	0.305295	0.322067	0.838983	00:10
8	0.295115	0.318580	0.843220	00:10
9	0.294310	0.314723	0.847458	00:10

```
CPU times: user 1min 42s, sys: 626 ms, total: 1min 43s
Wall time: 1min 43s
```

Figure 8.30 – Results from a 10-epoch run training a model with tabular data

The goal of this training process is to get a model with the best accuracy. With this goal in mind, there are two problems with the results shown in *Figure 8.30*:

1. The best accuracy is in epoch 5, but the model we get at the end of the training process has the accuracy from the last epoch. The output of `validate()` for this model, shown in *Figure 8.31*, shows that the accuracy for the model is not the best accuracy from the training run:

    (#2) [0.3147226572036743,0.8474576473236084]

    Figure 8.31 – Output of validate() for the model trained with the 10-epoch run

    Note that if the accuracy was still steadily improving up to the end of the training run, the model would likely not be overfitting yet, though you would want to validate this by exercising the trained model with the test set.

2. The training run keeps going after the accuracy is no longer improving, so training capacity is being wasted. For a run like this one, where every epoch completes in less than a minute, it isn't really a problem. However, consider the impact of wasted training cycles if each epoch took an hour and you were running on a paid Gradient instance. In that case, you could end up wasting many dollars on training cycles that didn't improve the performance of the model.

Luckily, fastai includes a solution to both of these problems: **callbacks**. You can use callbacks to control the training process and automatically take actions during training. In this recipe, you will learn how to specify callbacks in fastai that stop the training process when the model isn't getting any better and save the best model from the training run. You will revisit the model you trained in the *Training a model in fastai with a curated tabular dataset* recipe of *Chapter 3, Training Models with Tabular Data*, but this time you will control the training process for the model using fastai callbacks.

Getting ready

Confirm that you can open the `training_with_tabular_datasets_callbacks.ipynb` notebook in the `ch8` directory of your repo.

How to do it...

In this recipe, you will be running through the `training_with_tabular_datasets_callbacks.ipynb` notebook. In this recipe, you will train three variations of the model:

- Training with no callbacks

- Training with a single callback to end the training process when it has stopped improving

- Training with two callbacks: one to end the training process when it has stopped improving, and one to save the best model

Once you have the notebook open in your fastai environment, complete the following steps:

1. Update the following cell to ensure that `model_path` points to a writeable directory in your Gradient or Colab instance:

    ```
    model_path = '/notebooks/temp'
    ```

2. Run the cells in the notebook up to the `Define and train the model with no callbacks` cell to import the required libraries, set up your notebook, and define a `dataloaders` object for the `ADULT_SAMPLE` dataset.

    By calling `set_seed()` this way and defining a `dataloaders` object before each training run, we can get consistent training results across multiple training runs. You will see the significance of having consistent training results when we compare the results for training with and without callbacks.

3. Run the following cell to define and train a model with no callbacks:

    ```
    %%time
    set_seed(dls,x=42)
    learn = tabular_learner(dls,layers=[200,100],
    metrics=accuracy)
    learn.fit_one_cycle(10)
    ```

    The call to `set_seed()` specifies that the random seeds related to the model are set to `42` (an arbitrary value) and that results are reproducible, which is exactly what we want.

The output of this cell shows the training loss, validation loss, and accuracy by epoch, as well as the overall time taken for the training run, as shown in *Figure 8.32*. Note that the training run goes for all 10 epochs, even though the accuracy stops improving after epoch 2:

epoch	train_loss	valid_loss	accuracy	time
0	0.339484	0.361489	0.822034	00:08
1	0.339429	0.366499	0.826271	00:07
2	0.323125	0.321370	0.864407	00:06
3	0.325409	0.342183	0.838983	00:06
4	0.326003	0.335209	0.843220	00:06
5	0.317508	0.336868	0.830508	00:06
6	0.310379	0.330359	0.830508	00:06
7	0.297920	0.334773	0.843220	00:06
8	0.297238	0.333442	0.843220	00:06
9	0.282386	0.332747	0.843220	00:06

```
CPU times: user 1min 6s, sys: 499 ms, total: 1min 7s
Wall time: 1min 8s
```

Figure 8.32 – Results of training a model with no callbacks

4. Run the following cell to get the accuracy for the trained model:

```
learn.validate()
```

The output of this cell is shown in *Figure 8.33*. The accuracy for the model is the accuracy achieved in epoch 9, the final epoch of the training run:

```
(#2) [0.3327472507953644,0.8432203531265259]
```

Figure 8.33 – Output of validate() for a model trained with no callbacks

5. Run the following cell to define and train a model with one callback – early stopping when the model's accuracy no longer improves:

```
%%time
set_seed(dls,x=42)
learn_es = tabular_learner(dls,layers=[200,100],
metrics=accuracy)
learn_es.fit_one_
cycle(10,cbs=EarlyStoppingCallback(monitor='accuracy',
min_delta=0.01, patience=3))
```

The `fit` statement for this model includes the definition of an `EarlyStoppingCallback` callback. Here are the arguments used to define the callback:

a) `monitor='accuracy'` – specifies that the callback is tracking the value of the accuracy metric. When the accuracy metric stops improving for the specified period, the callback will be triggered.

b) `min_delta=0.01` – specifies that the callback pays attention to changes in accuracy between epochs that are at least 0.01. That is, if the change in accuracy between epochs is less than 0.01, the callback ignores the change.

c) `patience=3` – specifies that the callback is triggered when the accuracy metric stops improving for 3 epochs.

The output of this cell shows the training loss, validation loss, and accuracy by epoch, as well as the overall time taken for the training run, as shown in *Figure 8.34*:

epoch	train_loss	valid_loss	accuracy	time
0	0.339484	0.361489	0.822034	00:07
1	0.339429	0.366499	0.826271	00:09
2	0.323125	0.321370	0.864407	00:07
3	0.325409	0.342183	0.838983	00:08
4	0.326003	0.335209	0.843220	00:08
5	0.317508	0.336868	0.830508	00:08

```
No improvement since epoch 2: early stopping
CPU times: user 50.1 s, sys: 326 ms, total: 50.4 s
Wall time: 50.5 s
```

Figure 8.34 – Results of training a model with an early stopping callback

Note that now the training stops after epoch 5. You can see that the metric being tracked by the callback, `accuracy`, stops improving after epoch 2. Once 3 more epochs have passed (the value set for the `patience` parameter for the callback), the training process stops early, at epoch 5, even though the `fit_one_cycle()` call specifies a run of 10 epochs. So, with an early stopping callback, we save 4 epochs and about 25% of the training time (51 seconds with the early stopping callback versus 68 seconds with no callbacks).

6. Run the following cell to get the accuracy for the trained model:

```
learn_es.validate()
```

The output of this cell is shown in *Figure 8.35*. The accuracy for the model is the accuracy achieved in epoch 5, the final epoch of the training run:

(#2) [0.33686837553977966,0.8305084705352783]

Figure 8.35 – Output of validate() for a model trained with an early stopping callback

7. With an early stopping callback, we reduce the training time compared to a model with no callbacks, but the accuracy of the trained model is lower than the best accuracy achieved during the training run. Let's train another model that includes a callback to save the best model. This callback will ensure that the accuracy of the trained model is the best result we get from the training run. To do this, start by running the following cell to define and train a model with two callbacks – early stopping when the model's accuracy no longer improves, and saving the best model:

```
%%time
set_seed(dls,x=42)
learn_es_sm = tabular_learner(dls,layers=[200,100],
metrics=accuracy)
keep_path = learn_es_sm.path
# set the model path to a writeable directory. If you
don't do this, the code will produce an error on Gradient
#learn_es_sm.path = Path('/notebooks/temp/models')
learn_es_sm.path = Path(model_path)
learn_es_sm.fit_one_
cycle(10,cbs=[EarlyStoppingCallback(monitor='accuracy',
min_delta=0.01,
patience=3),SaveModelCallback(monitor='accuracy', min_
delta=0.01)])
# reset the model path
learn_es_sm.path = keep_path
```

In addition to the definition of an `EarlyStoppingCallback` callback like
the one specified in *Step 5*, the `fit` statement for this model also includes
a `SaveModelCallback` callback. Here the arguments used to define this callback
are as follows:

a) `monitor='accuracy'` – specifies that the `SaveModelCallback` callback
   is tracking the value of the accuracy metric. The model will be saved for epochs
   where the accuracy hits a new high-water mark.

b) `min_delta=0.01` – specifies that the callback pays attention to changes in
   accuracy between epochs that are at least 0.01.

Note that this cell also includes statements to adjust the model path to a directory
that is writeable in Gradient. If you don't change the model's path to a writeable
directory, you will get an error in Gradient when you run this cell. Also, note that
you may see the **Saved filed doesn't contain an optimizer state** warning message
when you run this cell – you don't need to worry about this message for the
purposes of this recipe.

The output of this cell shows the training loss, validation loss, and accuracy by
epoch, as well as the overall time taken for the training run, as shown in *Figure 8.36*.
Note that the training stops after epoch 5 thanks to the early stopping callback:

epoch	train_loss	valid_loss	accuracy	time
0	0.339484	0.361489	0.822034	00:08
1	0.339429	0.366499	0.826271	00:09
2	0.323125	0.321370	0.864407	00:09
3	0.325409	0.342183	0.838983	00:09
4	0.326003	0.335209	0.843220	00:09
5	0.317508	0.336868	0.830508	00:09

```
Better model found at epoch 0 with accuracy value: 0.8220338821411133.
Better model found at epoch 2 with accuracy value: 0.8644067645072937.
No improvement since epoch 2: early stopping
CPU times: user 54.6 s, sys: 300 ms, total: 54.9 s
Wall time: 55 s
```

Figure 8.36 – Results of training a model with an early stopping callback and a model saving callback

8. Run the following cell to get the accuracy for the trained model:

```
learn_es.validate()
```

The output of this cell is shown in *Figure 8.37*. The accuracy for the model is the accuracy achieved in epoch 2, the best accuracy during the training run:

$$(\#2)\ [0.32137033343315125, 0.8644067645072937]$$

Figure 8.37 – Output of validate() for a model trained with early stopping and model saving callbacks

With both callbacks, we avoid running epochs that won't improve the performance of the model and we end up with a trained model with the best performance out of the epochs of the training run.

Congratulations! You have successfully applied callbacks in fastai to optimize the training process so that you get the most out of the training cycle for your models.

How it works...

This recipe demonstrated how you can use fastai callbacks to control the training process so you get the best results from the system capacity and time that you apply to training. There are a few additional details about fastai callbacks that are worth reviewing. In this section of the recipe, we will dig a bit deeper into how callbacks are used in fastai.

The set_seed() function is not the default function for fastai

In order to get clear comparisons between the performance of the model with and without callbacks, we needed to control random differences between training runs. That is, we want to train the model multiple times and get consistent loss and accuracy for epochs between training runs. For example, if the accuracy reported in epoch 1 is 0.826271 for the first training run, we want to get exactly the same accuracy in epoch 1 for each subsequent training run. By ensuring consistent performance between training runs, we can do *apples-to-apples* comparisons between runs, focusing on the impact of the callbacks rather than random differences between runs.

In this recipe, we used the `set_seed()` function to control random differences between training runs. You may have noticed that while fastai documentation includes a `set_seed()` function (`https://docs.fast.ai/torch_core.html#set_seed`), we don't use that function in the recipe. Instead, we use the following function, which is adapted from code shared in this forum discussion – `https://github.com/fastai/fastai/issues/2832`:

```
def set_seed(dls,x=42):
    random.seed(x)
    dls.rng.seed(x)
    np.random.seed(x)
    torch.manual_seed(x)
    torch.backends.cudnn.deterministic = True
    torch.backends.cudnn.benchmark = False
    if torch.cuda.is_available():
        torch.cuda.manual_seed_all(x)
```

Why use this custom `set_seed()` function instead of the default `set_seed()` function documented by fastai? The simple reason is that the default `set_seed()` function did not work as expected – I didn't get consistent training results with it. With the `set_seed()` function listed previously, on the other hand, I was able to get consistent training results.

The Callbacks section of summary() doesn't include the callbacks defined in the recipe

You may notice that the end of the `training_with_tabular_datasets_callbacks.ipynb` notebook includes calls to the `summary()` function for the learner objects for the three models that you trained in the recipe. You might expect the `Callbacks` section of the output of the `summary()` function for the two models that include callbacks, `learn_es` and `learn_es_sm`, would list the early stopping and model saving callbacks, but that is not the case. *Figure 8.38* shows the `Callbacks` section for the two models that have explicitly defined callbacks, and this section is identical to the `Callbacks` section for the model with no callbacks:

```
Callbacks:
  - TrainEvalCallback
  - Recorder
  - ProgressCallback
```

Figure 8.38 – Callbacks section of summary() output

Why doesn't the `Callbacks` section of `summary()` output include the callbacks defined with the model? The `summary()` function just lists the callbacks defined by fastai itself, not the callbacks that you define.

Is there anything else you can do with callbacks in fastai?

In this recipe, we used callbacks to ensure that the training cycle was as efficient as possible, but that's just scratching the surface of what you can do with callbacks in fastai. The fastai framework supports a broad range of callbacks to control aspects of the training process. In fact, in this book you have come across several kinds of fastai callbacks:

- **Tracking callbacks** – the early stopping and model saving callbacks that we used in this recipe are examples of tracking callbacks. This category of callbacks is documented here: `https://docs.fast.ai/callback.tracker.html`.

- **Progress and logging callbacks** – you saw an example of a callback in this category in the *Getting more details about image classification models* recipe, where you used a `ShowGraphCallback` callback to display a graph of training and validation loss. Progress and logging callbacks are documented here: `https://docs.fast.ai/callback.progress.html`.

- **Callbacks for mixed-precision training** – in the *Training a deep learning language model with a curated text dataset* recipe of *Chapter 4, Training Models with Text Data*, you used `to_fp16()` to specify mixed-precision training for the language model you trained in that section. Callbacks for mixed-precision training are documented here: `https://docs.fast.ai/callback.fp16.html`.

The callbacks that you have used so far by working through the recipes in this book show you some of the power and flexibility that you can get by using callbacks with fastai.

Making your model deployments available to others

In *Chapter 7, Deployment and Model Maintenance*, you deployed a couple of fastai models. To get a prediction, you pointed your browser to `localhost:5000` and that opened up `home.html` where you set your scoring parameters, requested a prediction, and then got a prediction back in `show-prediction.html`. All this happened on your local system. Through the web deployments done in *Chapter 7, Deployment and Model Maintenance*, you can only get to the deployment on your local system because `localhost` is only accessible on your local system. What if you wanted to share these deployments with friends to allow them to try out your models on their own computers?

The simplest way to do this is using a tool called **ngrok** that lets you share `localhost` on your computer with people working on other computers. In this recipe, we will go through steps that show you how to use ngrok to make your deployments available to others.

Getting ready

Follow the instructions at `https://dashboard.ngrok.com/get-started` to set up a free ngrok account and to set up ngrok on your local system. Note the directory where you install ngrok – you will need it in the recipe.

How to do it...

With the help of ngrok, you can get a URL you can share with your friends so they can try out your fastai model deployments. This recipe will show you how to share your deployment of the tabular model.

To share the deployment you created in the *Deploying a fastai model trained on a tabular dataset* recipe of *Chapter 7, Deployment and Model Maintenance*, go through the following steps:

1. Make the directory where you installed ngrok your current directory.

2. Enter the following command on the command line/terminal:

    ```
    .\ngrok http 5000
    ```

    This command produces output as shown in *Figure 8.39*. Copy the `https` `Forwarding` URL:

```
ngrok by @inconshreveable

Session Status          online
Account                 Mark Ryan (Plan: Free)
Update                  update available (version 2.3.40, Ctrl-U to update)
Version                 2.3.34
Region                  United States (us)
Web Interface           http://127.0.0.1:4040
Forwarding              http://aefa693e9059.ngrok.io -> http://localhost:5000
Forwarding              https://aefa693e9059.ngrok.io -> http://localhost:5000

Connections             ttl     opn     rt1     rt5     p50     p90
                        0       0       0.00    0.00    0.00    0.00
```

Figure 8.39 – Output of ngrok

3. Start the deployment of the tabular model by making `deploy_tabular` your current directory and entering the following command:

```
python web_flask_deploy.py
```

4. On a computer other than your local system, point the browser at the `https Forwarding` URL that you copied in *Step 2*. You should see `home.html`, as shown in *Figure 8.40*:

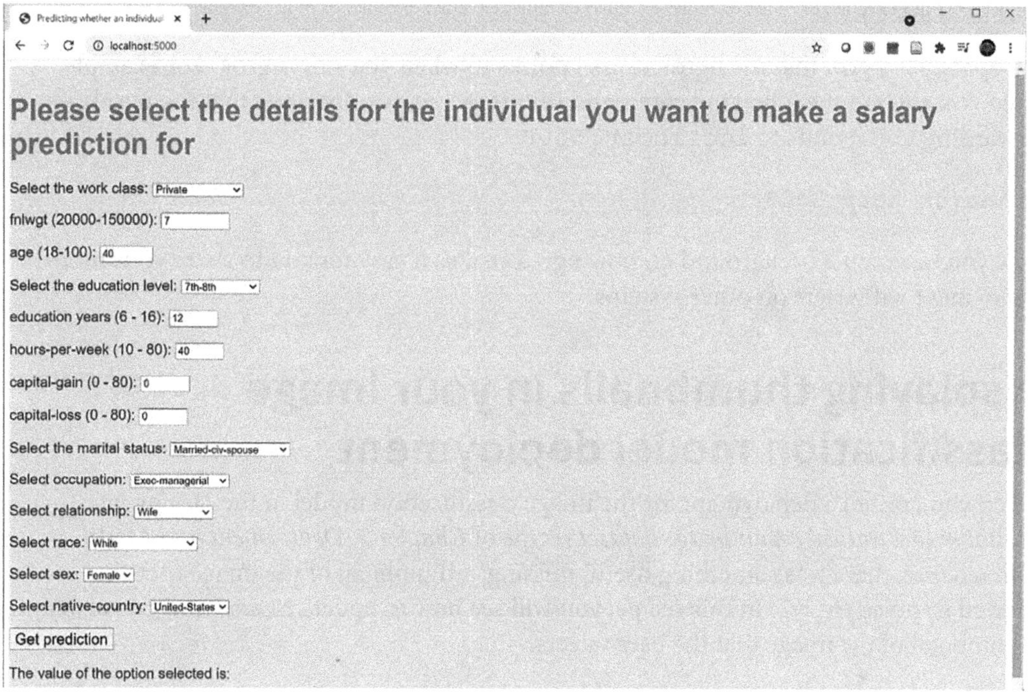

Figure 8.40 – home.html

5. In `home.html`, select **Get prediction** and confirm that you see a prediction in `show-prediction.html`, as shown in *Figure 8.41*:

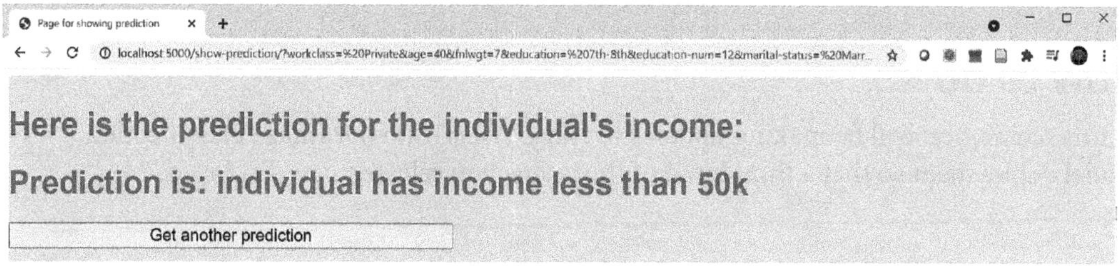

Figure 8.41 – show-prediction.html

Congratulations! You have successfully shared one of your fastai model deployments so that it is available to users on other systems with whom you share the ngrok forwarding URL.

How it works...

When you run ngrok, you create a secure tunnel to `localhost` on your local system. When you share the forwarding URL that is returned by ngrok, the people who receive the URL can point their browser to that URL to see what is being served at `localhost` on your local system.

You specify the port that the ngrok tunnel points to when you start ngrok. For example, when you entered the following command in this recipe, you specified that the ngrok forwarding URL points to `localhost:5000`:

```
.\ngrok http 5000
```

Now you have some background on how ngrok makes it easy for you to share your model deployment with users on other systems.

Displaying thumbnails in your image classification model deployment

When you created a deployment for the image classification model in the *Deploying a fastai model trained on an image dataset* recipe of *Chapter 7, Deployment and Model Maintenance*, there was something useful missing: a thumbnail of the image that you selected in `home.html`. In this recipe, you will see how to update `home.html` to display a thumbnail of the image that the user selects.

Getting ready

Ensure that you have followed the steps in the *Deploying a fastai model trained on an image dataset* recipe of *Chapter 7, Deployment and Model Maintenance*, to do the deployment of the image classification model.

How to do it...

In this recipe, you will be making updates to `home.html` in your image classification model deployment so that a thumbnail of the image you selected gets displayed.

To make these updates, go through the following steps:

1. Make the directory for the image classification deployment, `deploy_image`, your current directory.

2. Make the `templates` subdirectory your current directory by running the following command:

```
cd templates
```

3. Open `home.html` in an editor. I like to use Notepad++ (see `https://notepad-plus-plus.org/`), but you can use the editor of your choice.

4. Update the control for the file dialog to specify an `onchange` action of calling the `getFile()` JavaScript function. After the update, the control for the file dialog should look like this:

```
<input type="file"
       id="image_field" name="image_field"
       accept="image/png, image/jpeg"
      onchange="getFile();">
```

5. Define a new JavaScript function in `home.html` called `getFile()`:

```
<script>
function getFile() {
    file_list = document.getElementById("image_field").
files;
    img_f = document.createElement("img");
    img_f.setAttribute("id","displayImage");
    img_f.setAttribute("style","width:50px");
    img_f.setAttribute("alt","image to display here");
    document.body.appendChild(img_f);
    document.getElementById("displayImage").src = \
URL.createObjectURL(file_list[0]);
 }
</script>
```

Here are the key items in this function definition:

a) `file_list = document.getElementById("image_field).files;` – specifies that `file_list` contains the list of files associated with the `image_field` file selector

b) `img_f = document.createElement("img");` – defines a new image element called `img_f` on the page

c) `img_f.setAttribute("id","displayImage");` – assigns the `displayImage` ID to the new image element

d) `document.body.appendChild(img_f);` – adds the new image element to the bottom of the page

e) `document.getElementById("displayImage").src = URL.createObjectURL(file_list[0]);` – specifies that the file selected in the file dialog is displayed in the new image element

6. Save your changes to `home.html` and make `deploy_image` your current directory by running the following command:

    ```
    cd ..
    ```

7. Start the Flask server by running the following command:

    ```
    python web_flask_deploy_image_model.py
    ```

8. Open `localhost:5000` in your browser to display `home.html`.

9. Select the **Choose File** button and then select an image file from the `test_images` directory. If everything works, you should see a thumbnail of the image you selected at the bottom of the page, as shown in *Figure 8.42*:

Please select the image you want to classify:

Please choose images: [Choose File] 5_100.jpg

[Get prediction]

Figure 8.42 – home.html with a thumbnail of the selected image displayed

Congratulations! You have updated the deployment for the image classification model so that you see a thumbnail when you select an image.

How it works...

In this recipe, you have seen a small example of dynamically creating an element in an HTML page. Unlike all the other elements in home.html, the image element where we show the thumbnail is not there when home.html is first loaded. The image element only gets created when an image file has been selected and the getFile() function gets called. Why didn't we just have the image element hardcoded to be there when the file is first loaded, like the other controls?

If we had hardcoded the image element, then when you loaded home.html, before an image file had been selected, you would see an ugly missing-image graphic where the thumbnail should be, as shown in *Figure 8.43*:

Please select the image you want to classify:

Please choose images: [Choose File] No file chosen

[Get prediction] ⊠image to display here

Figure 8.43 – home.html with a hardcoded image element

By dynamically creating the image element only after an image had been selected, we can avoid the messy missing-image graphic.

You may remember that in the *Maintaining your fastai model* recipe of *Chapter 7, Deployment and Model Maintenance*, I mentioned that in a professional deployment you would not have to manually adjust the controls in home.html when the dataset schema got new values for categorical columns or brand-new columns. You could use the technique described in this recipe, dynamically creating controls, for most of the controls in home.html to make the deployment easier to adapt to schema changes.

Test your knowledge

In this chapter, we have reviewed a broad range of topics, from taking full advantage of the information that fastai provides about models to making your web deployments available to users outside of your local system. In this section, you will get the opportunity to exercise some of the concepts you learned about in this chapter.

Explore the value of repeatable results

In the *Using callbacks to get the most out of your training cycle* recipe, you made a call to the set_seed() function prior to training each of the models. In that recipe, I stated that these calls were necessary to ensure repeatable results for multiple training cycles. Test out this assertion yourself by following these steps:

1. First, make a copy of the training_with_tabular_datasets_callbacks. ipynb notebook.

2. Update your new notebook by commenting out the first call to set_seed() and rerun the whole notebook. What differences do you see in the output of fit_ one_cycle() between the model with no callbacks and the model with an early stopping callback? What about differences in the output of fit_one_cycle() between the model with an early stopping callback and the model with both an early stopping and a model saving callback?

3. Update your new notebook again by commenting out the second call to set_ seed() and rerun the whole notebook. Now what differences do you see in the output of fit_one_cycle() between the model with no callbacks and the model with an early stopping callback? What about differences in the output of fit_one_ cycle() between the model with an early stopping callback and the model with both an early stopping and a model saving callback?

4. Finally, update your notebook again by commenting out the final call to set_ seed() and rerun the whole notebook again. Compare the results of the models again. Has anything changed?

Congratulations! Hopefully, by following these steps, you have been able to prove to yourself the value of repeatable results. When you are comparing different variations of a model and you want to ensure you are getting an *apples-to-apples* comparison between variations, it can be very valuable to use the facilities in fastai to control the random initialization of aspects of the model so that you are guaranteed consistent results from multiple training runs.

Displaying multiple thumbnails in your image classification model deployment

In the *Displaying thumbnails in your image classification model deployment* recipe, you learned how to enhance the image classification deployment from the *Deploying a fastai model trained on an image dataset* recipe of *Chapter 7, Deployment and Model Maintenance*, by showing a thumbnail of the image selected for classification. You may recall that in the *Test your knowledge* section of *Chapter 7, Deployment and Model Maintenance*, you went through the exercise of enhancing the image classification model deployment by allowing the user to select multiple images for classification. What if you want to combine these two enhancements by allowing the user to select multiple images for classification and showing thumbnails of each selected image? Go through the following steps to explore how you would do this:

1. Ensure you have completed the exercise in the *Test your knowledge* section of *Chapter 7, Deployment and Model Maintenance*, to create a deployment of the image classification model that allows the user to select multiple images to be classified. You will be updating the code you completed in that section so that it shows thumbnails of the selected images.

2. Make a copy, called `deploy_image_multi_test`, of the `deploy_image_test` directory where you created the multi-image classification deployment. To do this, make the directory that contains `deploy_image_test` your current directory and run the following command:

    ```
    cp -r deploy_image_test deploy_image_multi_test
    ```

3. Make `deploy_image_multi_test/templates` your current directory. You will be making updates to the `home.html` file in this directory.

4. In `home.html`, update the control for the file dialog to specify an `onchange` action of calling the `getFile()` JavaScript function. After the update, the control for the file dialog should look like this:

    ```
    <input type="file" multiple
            id="image_field" name="image_field"
            accept="image/png, image/jpeg"
        onchange="getFile();">
    ```

5. Define a new JavaScript function in `home.html` called `getFile()`. This function will be a generalization of the `getFile()` function you defined in the *Displaying thumbnails in your image classification model deployment* recipe:

```
function getFile() {
   img_f = [];
   var i = 0;
   var di_string = "displayImage"
   file_list = \
document.getElementById("image_field").files;
   for (file_item of file_list) {
      img_f[i] = document.createElement("img");
      var di_1 = di_string.concat(i)
      img_f[i].setAttribute("id",di_1);
      img_f[i].setAttribute("style","width:50px");
      img_f[i].setAttribute("alt","image to display here");
      document.body.appendChild(img_f[i]);
      document.getElementById(di_1).src =\
URL.createObjectURL(file_item);
      i =  i+1
   }
}
```

Here are the key items in this function definition:

a) `file_list = document.getElementById("image_field").files;` – specifies that `file_list` contains the list of files associated with the `image_field` file selector.

b) `var di_string = "displayImage"` – defines the `di_string` string that will be used as the prefix of the IDs of the image elements that will be added to the page.

c) `for (file_item of file_list)` – specifies that the `for` loop iterates through the items in `file_list`. For each item in `file_list`, an image element will be created to display the image associated with the item.

d) `img_f[i] = document.createElement("img");` – defines a new `img_f[i]` image element on the page.

e) var di_1 = di_string.concat(i) – defines a di_1 string using the
dl_string prefix and the i index. For example, the first time through the loop,
the value of di_1 will be displayImage1.

f) img_f[i].setAttribute("id",di_1); – assigns the di_1 ID to the
img_f[i] image element.

g) document.body.appendChild(img_f[i]); – adds the img_f[i] image
element to the bottom of the page.

h) document.getElementById(di_1).src = URL.
createObjectURL(file_item); – specifies that the image file associated
with file_item is displayed in the img_f[i] image element.

With these changes, thumbnails for the files that the user selects in the file dialog
will be displayed at the bottom of home.html.

6. Now test whether everything works. Save your changes to home.html, make
deploy_image_multi_test your current directory, and start the Flask server
by running the following command:

```
python web_flask_deploy_image_model.py
```

7. Open localhost:5000 in your browser to display home.html.

8. Select the **Choose Files** button and then select image files from the test_images
directory. If everything works, you should see thumbnails of each of the images you
selected at the bottom of the page, as shown in *Figure 8.44*:

Please select the images you want to classify:

Please choose images: [Choose Files] 3 files

[Get prediction]

Figure 8.44 – home.html showing thumbnails for multiple selected images

Congratulations! You have combined two enhancements to the image classification
deployment to allow your users to select multiple images for the model to classify and see
thumbnails for the images they selected.

Conclusion and additional resources on fastai

In this book, you have explored a broad range of the capabilities of the fastai framework. By adapting the recipes in this book, you should be able to apply fastai to create deep learning models to make predictions on a wide variety of datasets. You will also be able to deploy your models in simple web applications.

There are many more capabilities in fastai beyond those covered in this book. Here are some additional fastai resources that you can use to learn more about the platform:

- To dig deeper into fastai, you can check out the online documentation for the framework (`https://docs.fast.ai/`).

- If you want a comprehensive guide to fastai, I highly recommend the outstanding book from Jeremy Howard (the originator of fastai) and Sylvain Gugger *Deep Learning for Coders with Fastai and PyTorch* (`https://www.amazon.com/Deep-Learning-Coders-fastai-PyTorch/dp/1492045527`).

- Jeremy Howard's YouTube channel (`https://www.youtube.com/user/howardjeremyp`) is another excellent source of information about fastai, including videos of Howard's deep learning course built on fastai, *Practical Deep Learning for Coders* (`https://course.fast.ai/`).

- When you are ready to go even deeper, Zachary Mueller's *Walk with fastai* site (`https://walkwithfastai.com/`) is a fantastic resource that consolidates many insights from the fastai forum (`https://forums.fast.ai/`) along with Mueller's own encyclopedic understanding of the platform.

Thank you for taking the time to read this book and following the recipes in it. I hope that you have found this book useful, and I encourage you to apply what you have learned to do great things by using fastai to harness the power of deep learning.

Other Books You May Enjoy

If you enjoyed this book, you may be interested in these other books by Packt:

Mastering spaCy

Duygu Altinok

ISBN: 978-1-80056-335-3

- Install spaCy, get started easily, and write your first Python script
- Understand core linguistic operations of spaCy
- Discover how to combine rule-based components with spaCy statistical models
- Become well-versed with named entity and keyword extraction
- Build your own ML pipelines using spaCy
- Apply all the knowledge you've gained to design a chatbot using spaCy

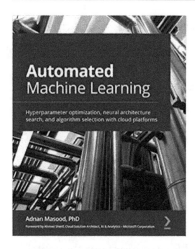

Automated Machine Learning

Adnan Masood

ISBN: 978-1-80056-768-9

- Explore AutoML fundamentals, underlying methods, and techniques
- Assess AutoML aspects such as algorithm selection, auto featurization, and hyperparameter tuning in an applied scenario
- Find out the difference between cloud and operations support systems (OSS)
- Implement AutoML in enterprise cloud to deploy ML models and pipelines
- Build explainable AutoML pipelines with transparency
- Understand automated feature engineering and time series forecasting
- Automate data science modeling tasks to implement ML solutions easily and focus on more complex problems

Packt is searching for authors like you

If you're interested in becoming an author for Packt, please visit authors. packtpub.com and apply today. We have worked with thousands of developers and tech professionals, just like you, to help them share their insight with the global tech community. You can make a general application, apply for a specific hot topic that we are recruiting an author for, or submit your own idea.

Share Your Thoughts

Now you've finished *Deep Learning with fastai Cookbook*, we'd love to hear your thoughts! Scan the QR code below to go straight to the Amazon review page for this book and share your feedback or leave a review on the site that you purchased it from.

https://packt.link/r/1-800-20810-3

Your review is important to us and the tech community and will help us make sure we're delivering excellent quality content.

Index

text data
  about 27, 28
  examining, with fastai 55-59
text datasets
thumbnails
  displaying, in image classification
    model deployment 304-307
tracking callbacks 301
trained tabular model
  saving 101-103
transfer learning
  used, for training image
    classification model 182-184
  using, recipe 224

U

Uniform Resource Locator (URL) 20

V

value, of repeatable results
  exploring 308

W

web deployment
  code 240-246
  examples 246
web deployment, of fastai image
    classification model
  code 252-254
  usage, overview 250-252
web deployment, of fastai tabular model
  usage, overview 238-240

www.ingramcontent.com/pod-product-compliance
Lightning Source LLC
Chambersburg PA
CBHW062058050326
40690CB00016B/3134